GUIDE TO
PREHISTORIC
ASTRONOMY
IN THE SOUTHWEST

REVISED AND UPDATED

T0162259

GUIDE TO
PREHISTORIC
ASTRONOMY
IN THE SOUTHWEST

REVISED AND UPDATED

BY J. McKIM MALVILLE

BOWER
HOUSE

DENVER

Cover and text design by Rebecca Finkel
Cover photo: Full moon rise December 26, 2004, by Helen Richardson

Library of Congress Cataloging-in-Publication Data
Malville, J. McKim.
A guide to prehistoric astronomy in the Southwest / J. McKim Malville—Rev. and updated.
p. cm.
Rev. ed. of: Prehistoric astronomy in the Southwest / J. McKim Malville,
Claudia Putnam. c2003.
Includes bibliographical references and index.
ISBN 978-1-55566-414-5
1. Indian astronomy—Southwest, New. 2. Archaeoastronomy—Southwest, New.
3. Indians of North America—Southwest, New—Antiquities. 4. Southwest, New—Antiquities.
I. Malville, J. McKim. Prehistoric astronomy in the Southwest. II. Title.
E78.S7M135 2008
979'.01—dc22
                              2008003145

9   8   7   6

CONTENTS

PREFACE

I HOPE THIS GUIDE finds its way into the hands of many visitors to the prehistoric ruins of the Southwest who are interested in exploring the astronomy that was developed by the Ancestral Puebloans. Our goal is to seek out the sky watchers who lived among these ruins. Perhaps everyone was a sky watcher, carefully watching the changing positions of the sun and moon on their horizons. For us, as guests in this remarkably beautiful landscape, the mixture of old stones, old stars, and stunning vistas that we find in the Four Corners area can generate mystery and adventure comparable to that of a visit to Stonehenge or Newgrange. One doesn't need to go to Machu Picchu for enigmatic ruins or Egypt for tantalizing hints of great knowledge hidden in the mute rocks.

Since the last edition of this book in 1993, there has been a steady growth in our knowledge of astronomy practiced by the Ancestral Puebloans. Predictions that had been made in the last edition have been verified in the many annual returns of the solstice sun and, after 18.6 years, the return of the major standstill moon. New work in archaeology has been published on Chimney Rock, Chaco Canyon, Mesa Verde, Yellow Jacket, and Hovenweep, which reveals even more clearly how culture and astronomy were integrated. Astronomy can now be understood as an essential aspect of the whole culture of the Ancestral Pueblo peoples.

A Guide to Prehistoric Astronomy in the Southwest synthesizes astronomy, people, and culture. After an introduction to basic astronomy and archaeology, the book identifies specific places where one can view evidence for astronomical practices, as well as observing essentially the same sunrises that were observed by the Ancestral Puebloans 1,000 years ago.

Astronomy did not arrive in the area fully born, and the book shows how astronomy evolved with the practical and ceremonial needs of the people. Living quarters and ceremonial spaces started out in parallel to the heavens, probably due to ancient memories of migrations from the north.

Astronomical calendars were needed to organize periodic festivals and integrate the vast spaces of the Chacoan regional system. Cycles of the sun and moon were used for ceremonies involving sacred time.

We have located calendrical stations in Chaco Canyon that would have provided the dates for winter and summer solstice festivals in the canyon. We now understand how Chimney Rock could have functioned as a Puebloan version of the Greenwich Observatory, by providing calendrical dates through a long-distance signaling network. New interpretations are presented for the famous sun dagger and supernova pictographs in Chaco Canyon. During the last few decades of the thirteenth century, especially when the Great Drought set in, life in the Southwest became difficult and precarious, and we can see how astronomy changed just before abandonment of the area.

ACKNOWLEDGMENTS

THIS GUIDE TO PREHISTORIC RUINS OF THE SOUTHWEST is an outgrowth of the book coauthored with Claudia Putnam. I thank Claudia for her contributions to that book.

My first introduction to the possibilities of astronomy at Yellow Jacket was provided by Mark Neupert in the summer of 1986. Hardworking field research teams of which I've been a part, who have fought the gnats and sage with determination and good humor, have included Dr. Carol Ambruster, Ken Brownsberger, John Cater, John Jacobs, Annie Jones, Jean Kindig, Matt Moody, Rudy Poglitsh, Jennifer Ritter, Jim Walton, and Cindy Webb. I have received encouragement and advice from southwestern archaeologists Joe Ben Wheat, Frank Eddy, Jim Judge, and Steve Lekson. I thank the Wilson family and the Archaeological Conservancy for permission to work on their lands at Yellow Jacket. I am indebted to Jean Kindig for her many fine drawings throughout the book. During my work at Mesa Verde, Frank Occhipinti, Greg Munson, Aaron Kaye, Don Ross, and Kelley Shepherd provided much appreciated field assistance.

"We shall not cease from exploration.
And the end of all our exploring
will be to arrive where we started
and know the place for the first time."

—T. S. ELIOT

CHAPTER 1

THE ANCESTRAL
PUEBLOAN ASTRONOMER

FIGURE 1.1. Equinox Sunrise at Chimney Rock.

According to Hopi tradition a child is born of Mother Earth and Father Sun as well as human parents:

> When a child was born, his Corn Mother was placed beside him, where it was kept for twenty days, and during this period he was kept in darkness . . . Early on the morning of the twentieth day, the mother, holding the child in her left arm and the Corn Mother in her right hand and accompanied by her own mother—the child's grandmother—left the house and walked to the east. Then they stopped, facing east, and prayed silently, casting pinches of cornmeal to the rising sun. When the sun cleared the horizon the mother stepped forward, held up the child to the sun, and said, "Father Sun, this is your child."[1]

In April 1535 the young Spanish priest, Cristóbal de Molina, observed one of the last great solar festivals of the Inca:

> *They waited in deep silence for the sun to rise. As soon as the sunrise began they started to chant in splendid harmony and unison. They all stayed there chanting from the time the sun rose until it had completely set. As the sun was rising toward noon they continued to raise their voices, and from noon onwards they lowered them, keeping careful track of the sun's course. They allowed their voices to die away on purpose. And as the sun was sinking completely and disappearing from sight they made a great act of reverence, raising their hands and worshipping it in the deepest humility.* "[2]

Above the rim of the horizon the deep blue extends to the edge of space. For many people, the blue of the overarching sky means infinity, mystery, and power. The sky is peopled with gods: sky gods, sun gods, wind gods, rain gods. Gods threw thunderbolts, carried the sun from east

FIGURE 1.2. The Sacred Plaza of Machu Picchu where sunrise at June solstice and sunset at December solstice may have been celebrated.

to west, tried to protect it from hungry demons who wished to swallow it. The first astronomers of our planet attended to the needs of these deities. They were servants and priests to mystery and power.

The science of astronomy has taken many forms over the past 5,000 years, always strongly reflecting the culture in which it was embedded. An extraordinary variety of institutions and practices has arisen to support this science, the oldest of all, in vastly different environments around the world.

Today's astronomers are physicists who seek to understand the light of distant stars and galaxies in terms of fundamental physical phenomena that occur on the earth. But we do more, for astronomers explore the fabric of time. With telescopes as our shovels, we dig back to our origins using the ancestral photons of the universe. The cosmologists of today are asking the questions people asked for thousands and thousands of years, sitting around a campfire at night, as they wondered about the meaning of those flickering but eternal stars overhead and the fragile, transient life around them.

In the subfield of astronomy known as archaeoastronomy, the challenge is to understand those ancient sky watchers and to be able see the heavens through their eyes. The astronomers who preceded us knew their heavens and their stars: the Ancestral Puebloan sun watchers, the court astrologers of Beijing and Babylon, the astronomer kings of Copán and Palenque, the astronomer-architects of the great temples of India and the stone circles of Great Britain, and even the blood-encrusted astronomer-priests of the Aztec capital of Tenochtitlán.

FIGURE 1.3. Calendar Circle at Nabta Playa.

The oldest known set of stones oriented to the heavens was put in place by nomadic cattle herders of the Sahara Desert some 7,000 years ago. They built a ceremonial center at the edge of a seasonal lake, Nabta Playa, which they visited after summer monsoon rains brought water and food for the cattle. The area contains a circle of stones that indicates the directions of north and the rising sun on summer solstice (Figure 1.3). The direction to north may have been established by the shadow casting of a vertical stone or gnomon, perhaps a precursor to the Egyptian obelisks (Figure 1.4).

FIGURE 1.4. Gnomon at Nabta Playa.

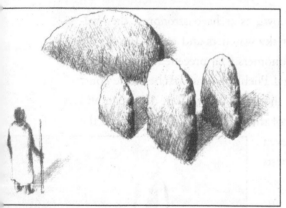

FIGURE 1.5. Megaliths of Nabta Playa.

The nomads also dragged large stones, weighing 1–2 tons, onto the lake bed and planted them in its sediments, aligning them with some of the brightest stars in their skies, such as Sirius, Arcturus, and Alpha Centauri (Figure 1.5).

Why did the nomads go to so much effort to build a ceremonial center that incorporated directions to the north, solstice sunrise, and bright stars? Their travels across the sands of the Sahara must have been guided by the stars in a manner similar to ancient mariners. Furthermore, it was at the time of June solstice that summer monsoons moved into the Sahara, bringing water to Nabta Playa. The important features of their lives, travel across the desert and water for themselves and their animals, were thus incorporated in the astronomy of their ceremonial center.[3]

These ancient astronomers served the needs of the societies in which they lived. In the millennia since Nabta Playa was abandoned to the sands, astronomers have advised emperors and generals, predicted eclipses and conjunctions of planets, devised calendars for festivals, and established dates for planting and harvesting. Sometimes overtly or sometimes subtly, these ancestral astronomers provided authority and legitimacy for their emperors and kings. Sometimes they were rewarded when their predictions were correct, other times they lost their heads when they were wrong.

Astronomers have used a variety of techniques for preserving and transmitting knowledge besides aligning megalithic stones. Today, in western science we have lectures, scientific journals such as the *Astrophysical Journal* and the British journal *Nature,* and meetings of the American Astronomical Society and the International Astronomical Union. In the past in addition to the written word, astronomical lore was transmitted through folk stories, myths, elaborate rituals and festivals, complex and symbolic architecture, dance, and song. The Ancestral Puebloan peoples have left us their sacred spaces, the organized rocks of architecture, and markings on canyon walls as records of their astronomy.

Those stones with which people have built their homes and temples can be transmitters of information about their innermost thoughts as well as information about the details of their astronomical knowledge. The structure of the temples of India, for example, provides insights into the beliefs of certain peoples concerning the nature and origin of their world. In their

FIGURE 1.6. The Sun Temple at Konarak, Orissa State, India. It is built in the form of a solar chariot with twenty-four wheels, drawn forward toward the east by seven horses.

stones are encoded the myths of innumerable gods and demons, many of whom were responsible in one way or another for the creation of the world.

Deep in their dark centers lies the *garbha griha,* the "womb chamber," symbolic of the powerful and dangerous chaos out of which our world developed. The center of the temple also represents a cave in the cosmic mountain, which itself was the axis of the world connecting heaven and earth. It is to this center that the devout Hindu returns periodically for worship and rejuvenation. The devotee will often enter the temple compound through a doorway opening to the east. In some temples of southern India, the rays of the rising sun at equinox sweep through long passageways into the sacred center and touch the image of the god enshrined therein.[4]

FIGURE 1.7. Templo Mayor. The twin shrines on its summit are to Tlaloc, the rain god (left), and Huitzilopochtli, the hummingbird god of sun and war (right).

FIGURE 1.8. The stone of Coyolxuahqui lies at the base of Templo Mayor.

The great temple, the Templo Mayor, at the center of the Aztec capital of Tenochtitlán, is another structure that informs us of the astronomical myths and the social customs of the Mexica.[5] The massive pyramid represents the primordial mountain of the Aztecs, the Serpent Mountain of Coatapec.

On its flanks the first battle between the forces of light and darkness took place. The god of the sun, Huitzilopochtli, was victorious, killing and dismembering the moon goddess, Coyolxuahqui. The stone of Coyolxuahqui, 11 feet across, once lay at the base of the Templo Mayor, near the Cathedral of Mexico City, bearing chilling witness to this awesome duel between the sun and moon (Figure 1.8). On that temple, with the great drum sounding and conch shells and horns blowing, that initial battle was reenacted in the form of human sacrifice. The Templo Mayor was oriented so that on the morning of the equinox, the sun would first appear between its twin towers.

The most comprehensive knowledge of prehistoric astronomy in the Southwest comes from the marvelous ruins of the Ancestral Puebloans, located at Mesa Verde, Chaco Canyon, Chimney Rock, Yellow Jacket, and Hovenweep. It is upon those places and the culture that they reveal that I focus my attention in this book. The people of the ancient Southwest lived close to the heavens and paid attention to what was overhead. The average Ancestral Puebloan probably had a better knowledge of the patterns of the stars and cycles of sun and moon than the average inhabitant of the Southwest today. A thousand years ago, people's lives were governed by the intertwining cycles of sun and moon, and observational astronomy must have played a major role in guiding ceremonial calendars.

Both the astronomical knowledge of the Ancestral Puebloans and the culture that gave it birth experienced great changes between 700 and 1300 CE, and in the following pages I will attempt to follow those six fascinating centuries of interaction between Puebloan culture and astronomical discovery. I identify four major themes in the evolution of Puebloan astronomy:

1. Ceremonies at the solstice. Since the sun dwells at these extreme points for several days, it is a straightforward task to establish horizon markers for solstice sunrises. Some farmers may have found it useful to know the date of summer solstice. However, in the Southwest, where rainfall can be very uncertain, there are far better ways to determine when to plant than using the position of the sun on the horizon.

2. North-south orientations. After 700 CE, individual households in the northern San Juan region oriented themselves along approximate north-south lines. As archaeologist William Lipe[6] observes, these households probably had considerable control over their own religious symbols and rituals, and these common north-south orientations displayed "spiritual independence." Each family may have established the direction to north on their own. That interest in north-south alignments and the ability to establish the north-south meridian may have been carried southward into Chaco Canyon during major migrations that occurred between approximately 870–920 CE.

During the next century, culminating in the Chaco fluorescence, local communities lost autonomy as ideological and political power appears to have influenced architectural design. Since some of the alignments of Chacoan great houses and kivas are much better than could have been achieved simply using the North Star, the technique they used was probably shadow-casting by a vertical pole, or gnomon. The orientations of large structures such as Casa Rinconada and Pueblo Bonito along north-south lines demonstrates both astronomical sophistication and an ability of leaders to design the buildings and to organize workers.

3. Major standstills of the moon. The discovery of the ability of the moon to travel beyond the limits of the sun probably came about because of the unusual geography at Chimney Rock, which displayed the motions of the moon in such a manner that the standstill cycle could not be ignored. As viewed from the high mesa of Chimney Rock, every 18.6 years the moon at major northern standstill rises between the chimneys, a feat that the sun can never achieve.

FIGURE 1.9. Moon rising between the rock towers of Chimney Rock at major standstill.

4. Unexpected events. For people who depended upon the regularity of the sun and moon for establishing the dates of periodic festivals and the changing seasons, it must have been unsettling and perhaps a little terrifying to encounter changes in the otherwise calm and predictable heavens. There were supernovas in 1006 and 1054 CE, which could not be ignored. The supernova of 1006 was the brightest exploding star ever viewed by humans. The more famous supernova of 1054 produced the spectacular Crab Nebula.

FIGURE 1.10. Crab Nebula photographed by the Hubble Space Telescope.

The eruption of Sunset Crater in Arizona in 1064 CE could have been seen from Chaco Canyon. The appearance of Halley's Comet in 1066 was so brilliant that it alarmed the Saxons before the Battle of Hastings.

FIGURE 1.11. Comet Halley 1986.

There was a total eclipse of the sun in 1097, which was the
only total solar eclipse visible in the canyon during the
Chacoan fluorescence. We can only speculate on the terror and
awe it generated.

Chaco Canyon was clearly the center of the first major spurt of astronomical activity. The Bonito Phase of Chaco Canyon, when most of the
majestic great houses were built, appears to have been such a time, when
the cycles of the heavens were perceived and utilized by sharp-eyed
observers. By providing dates for rituals, festivals, and pilgrimages, astronomical knowledge was one of the sources of social cohesion and integration that tied the far-flung Chacoan society together.[7] The Chacoan community at Chimney Rock may have provided important calendrical information to Chaco through a visual signaling network.[8] Other outliers may
have contributed to the growth of astronomical knowledge in ways we do
not yet understand. When Chaco Canyon's power began to fade in the
early years of the twelfth century, there appears to have been a movement
of leadership northward across the San Juan River to Aztec Ruin. By some
form of cultural osmosis, astronomical knowledge spread northward to
Mesa Verde and the many villages of the Great Sage Plain north of Cortez.
The populations of the major settlements in the area, such as Yellow
Jacket, Mesa Verde, and Hovenweep, must have inherited some of the
astronomical wisdom acquired at Chaco Canyon and Chimney Rock during the previous century.

CHAPTER 2

THE ROOTS OF
ASTRONOMY

stronomy didn't arrive in the Southwest by alien spacecraft,
although New Mexico has more than its fair share of strange lights
in the sky and UFO sightings. Neither is it likely that a detailed knowledge of the cycles of the sun and moon was carried to Chaco Canyon or
Mesa Verde by traders or astronomer-priests from Mesoamerica. The
Chacoans and cultures of Mexico may have shared some of their symbols,
but it is difficult to see how quantitative information about the cycles of
the sun and moon could have been carried northward. Most likely, the
astronomy of the Ancestral Puebloans was homegrown, the independent
invention of smart and alert people who paid attention to the heavens.
The remarkable burst of astronomical knowledge of the pueblos may have
occurred primarily during the cultural fluorescence that occurred in
Chaco Canyon during the eleventh century.

The aspects of the culture that appear to have stimulated and nur-
tured the growth of astronomy were migration, trade, periodic festivals,
pilgrimage, and, oddly enough, war, danger, and violence. Astronomy
may have dominated peoples' lives more than we can imagine. The sky
was a region of wonder and power, sunlight and rain, danger and suste-
nance. It was a useful ally, but also an uncertain provider. It provided rain,
sunshine, direction, and time. The sky contains the north celestial pole,
which keeps sailors and travelers in the Northern Hemisphere from wandering in circles. Combined with the ever-changing phases of the moon,
the sun can be used to establish a calendar, which allowed traders to establish dates for trade fairs and leaders to organize festivals. Finally, toward

the end of the thirteenth century, life became difficult due to uncertain rainfall, and there was an agonizing and poignant downward slide into warfare and violence. Towers were built at Hovenweep and Yellow Jacket for protection of families and water sources; people gathered into defensible villages such as Sand Canyon; massacres took place at Castle Rock. Judging from the astronomy in and around the towers of Hovenweep, people may have turned to the sun as a source of protection during unstable times.

The Power of the North

Astronomy as evidenced in architecture started north of the San Juan River. In 700 CE, people living in the areas north of the San Juan River lived in basic habitation units that consisted of an aboveground block of five to ten rooms; an early form of a kiva, or a "proto-kiva," to the south or southeast; and a trash pile, or midden, farther to the south or southeast.

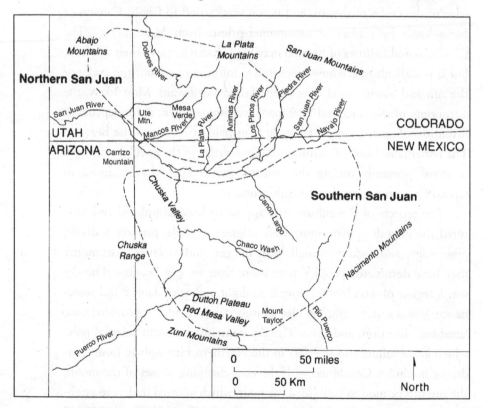

FIGURE 2.1. Northern and Southern San Juan Regions.

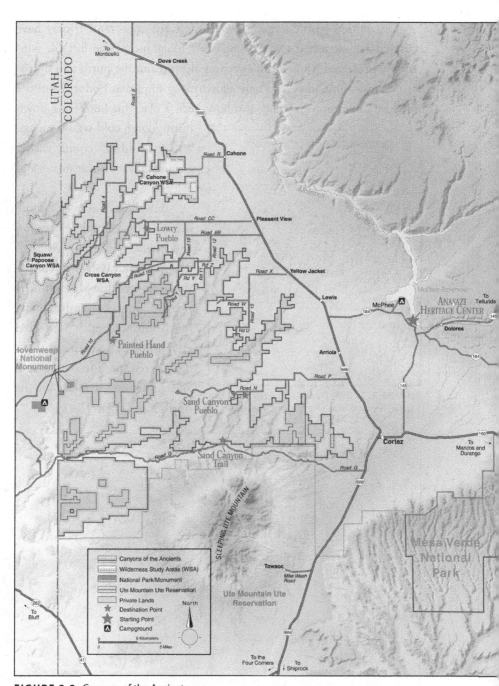

FIGURE 2.2. Canyons of the Ancients.

Orientation to the larger world was clearly important. It may have been partly practical: a south-facing room block meant comfortable winter days protected from winds sweeping down from the north and northwest. Those winds may also have blown away unpleasant odors from the midden. The proto-kiva may also have been a place for hunkering down during long, cold winter nights, where children could gather around elders to hear stories and legends.

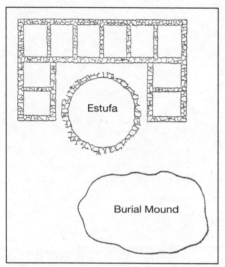

FIGURE 2.3. Basic Pueblo of the Northern San Juan with roomblock, kiva ("estufa"), and mid-

But these places had more meaning than simply protection and shelter from the weather. The proto-kivas had an intentional design that was symmetrical about an approximate north-south axis that was often established by a ventilator shaft, deflector stone, fire pit, sipapu, and sometimes a niche on the north wall (Figure 2.4).

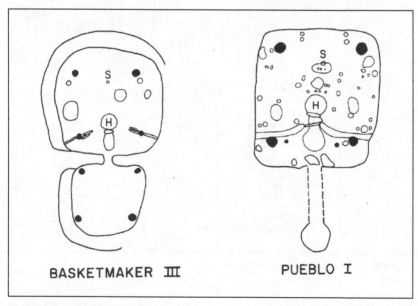

FIGURE 2.4. Basket Maker: Tres Bobos Hamlet (Axis = 162°). Pueblo I: Aldea Alfaeros (Axis = 153°). (H, hearth; S, sipapu).

There was more symbolism and purpose in these layouts than the direction of the wind from the north.

North was more of an abstract concept than it is for us living in the Northern Hemisphere today. We have Polaris in our sky, but in the eighth century there was no bright star near the North Pole. That empty place in the sky had meaning because stars revolved around it. That direction in space may have figured in legends about past migrations of ancestors from the north. People were outside at night looking to the north. Perhaps during the day they were watching the movement of shadows from a vertical pole and noting the direction of the shortest shadow.

Orientation of their homes must have been the result of a wish to align themselves in a proper and meaningful way to the larger landscape and the cosmos that lay beyond. This common astronomical theme has been described by Bill Lipe[1] as "rooted in beliefs and symbolic systems that were widely held through most of the San Juan Drainage and that long preceded the Chacoan fluorescence of ca. A.D. 1040–1135." The importance of north probably reveals deep memories of migrations of the past when their ancestors were making their way into North America.[2] Nomadic, seafaring, and migrating cultures typically develop knowledge of the cardinal directions and learn to read the sky. The nomadic pastoralists who moved with their herds across the Sahara of 7,000 years ago apparently knew their sky, as evidenced at Nabta Playa. Agricultural cultures pay attention to the patterns of the sun and especially the solstices. The Ancestral Puebloans had both migration and agriculture in their heritage, and it is not surprising that north was one significant feature of their cosmos.

These villages in the northern San Juan region were occupied for only a generation or two. People were moving around the landscape. During the early ninth century, the population in the northern San Juan increased to approximately 8,000 people. Fifty years later, there appear to have been a series of droughts from 880 to 900 CE, and the northern area was depopulated. Many of the people migrated southward to Chaco Canyon and the Chuska Mountains.

In the northern San Juan, individual households had a kind of spiritual independence; they controlled the symbolism that threaded their lives. The power and intensity of that spiritual symbolism is evidenced by the abandonment rituals that were practiced at some of the northern villages. Some, such as those of McPhee village, which was abandoned between

880 and 900, were vacated in a very systematic manner; the main ceremonial pit structure, or proto-kiva, was carefully burned, with ritual objects left in place and paired human burials nearby.[3] They moved away from the land but took with them some strongly held meaning and traditions.

Chaco Canyon was an attractive place to settle as the early 900s were years with above-average rainfall. The canyon provided protection from winter winds from the north. Successful maize farming depended on summer rains, which were probably reliable during the early 900s. During that time, agricultural surpluses were built up, and the first of the Great Houses were planned and built. The astronomy that developed may be traced to the changes in lifestyle and social structure following the migrations from north to south. Their settlements acquired a more permanent character; their independent spirituality and cosmologies were absorbed into a larger, more hierarchical society; and the canyon became a central place to visit for periodic festivals, feasting, and celebrations. All had a role in nurturing astronomy.

Pueblo Bonito was started about this time. It was a small and modest affair, much smaller, in fact, than the large villages of the northern San Juan (Figure 2.5). Its arch of rooms and central pit strongly resemble the

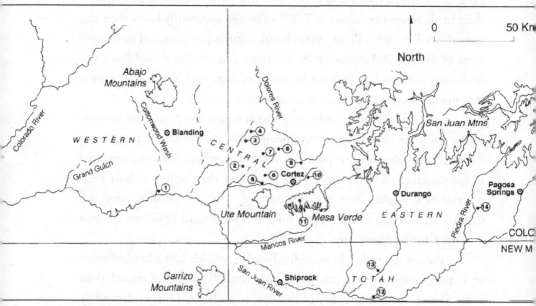

FIGURE 2.5. The Northern San Juan: 1) Bluff; 2) Seven Towers; 3) Lowry; 4) Ansel Hall; 5) Castle Rock; 6) San Canyon; 7) Albert Porter; 8) Yellow Jacket; 9) Escalante; 10) Wallace; 11) Farview; 12) Salmon; 13) Aztec; 14) Chimney Rock.

structure of the ceremonially abandoned northern villages. When they migrated to Chaco and were gathered up in a more inclusive society, the builders of the Great Houses took over these spatial conventions and made them more formal and accurate, linking themselves to ancient beliefs, traditions, and the cosmos.

Trade and the Calendar

Traders needed other traders as well as potential customers. In particular, they needed to arrange times at which to gather in Chaco. The easiest kind of calendar to establish among themselves involved the sun at its extreme northern or southern positions and the full moon. For example, at winter solstice they could arrange to meet at the full moon when the sun is farthest south. The full moon makes sense because the period leading up to it has nights of ever-increasing moonlight during which one can travel. A thousand years ago, people's lives were governed by the intertwining cycles of sun and moon, and observational astronomy must have played a major role in providing planting and ritual calendars.

It may all have started with traders, who converged on Chaco at those easily identifiable full moons at either summer or winter solstice. As the numbers of traders grew, the events developed into periodic festivals accompanying the trade. Perhaps one of the powerful men of Chaco, who lies buried in a crypt at Pueblo Bonito, organized the remarkable transformation of trade fairs into the Chaco phenomenon of periodic festivals, great houses, great kivas, and roads.

Periodic Festivals and Pilgrims

In the San Juan basin of northern New Mexico, Chaco Canyon appears to have served as the center of an extended regional system, with center-to-periphery distances of up to 155 miles and as many as 150 outlying communities. We identify a regional system as a network of geographically separate communities that are linked by a common ideology and by established trade networks. The distinguishing characteristics of the Chaco regional system are its Great Houses with their distinctive architecture, its great kivas, the network of roads, and implied sociopolitical complexity.

Ritual may have been the major link between the outliers and the central canyon, and residents of the Great Houses may have acquired power through control of the ritual calendar and ritual performance. Judge[4] proposed that the Great Houses were built to accommodate substantial numbers of individuals coming "on formal pilgrimages from the outlying areas to Chaco Canyon." Because Chaco Canyon had become a major turquoise processing center by around 1020 CE, trade and trade fairs may have been one of the reasons for the onset of a pilgrimage tradition. Because of the ritualistic function of turquoise, there may have developed a common symbiotic relationship between pilgrimage and trade.

Because the system of Chacoan outlying communities may have involved an area as large as 58,000 square miles, the practical issue of social integration over large distances is a genuine concern. It is hard to imagine a world without easy communication: no telephones, TV, telegraphs, or Pony Express riders. The remarkable road system centered on Chaco Canyon must trace out the arteries for the movement of traders and pilgrims. Sections of these roads are wider and more elaborate than the minimum necessary for foot travel, and suggest ritual movement such as groups of people walking in a parade-like fashion into the canyon or into a great house. Many temples in India, for example, open on to similarly wide streets, which are used for temple festivals. In the absence of a written language, the transfer of ideological knowledge, mythologies, and traditions can only occur through person-to-person contact at community gatherings such as pilgrimage events, festivals, or trade fairs.

Political power may not be needed to integrate a system such as Chaco's. Traders and pilgrims functioning as self-motivated individuals can provide substantial cultural cohesion.[5] Social order and cultural complexity can arise by means other than political persuasion, administrative control, or coercion. The cultural integration and societal transformations that appear to have been associated with Chaco can arise spontaneously. The essence of pilgrimage is the movement of people, individually or in groups, away from their homes to a sacred place as an act of religious devotion, and portions of the Chacoan road system may be manifestations of such voluntary human movement. Sections of the roads that are wide and carefully engineered are difficult to understand except as routes with ritual function and symbolic meaning. Human movement into the Canyon need not have been limited to the presently identified roads. Some sections of roads have been degraded by erosion, and other portions may have involved only rough trails. There are suggestions of as many as eight roads and road sections that converge upon Chaco Canyon.

The most clearly established Chacoan road is the North Road, which extends northward from Pueblo Alto to a stairway at the edge of Kutz Canyon (Figure 2.6). A pathway may once have continued down the heavily eroded floor of Kutz Canyon, past Twin Angels Pueblo to the San Juan River. Immediately across the river lies Salmon Pueblo. This great house may have been constructed at a convenient place to ford the river on the trail to Chaco.

The communities at Twin Angels, Halfway, and Pierre's, which would have been close to the North Road, were constructed in regions of poor soils unsatisfactory for agriculture. These houses make most sense as way stations for travelers. Between Aztec Ruin and

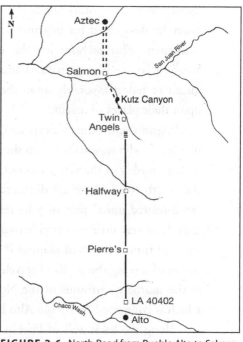

FIGURE 2.6. North Road from Pueblo Alto to Salmon.

Chaco they have an average spacing of about 18 miles, and they may have provided services for travelers and traders, such as water, cooking facilities, and shelter for the night.

The firepits on top of buttes and other features of Pierre's complex suggest activities associated with ideologically motivated movement along the northern road. Like the traditional bonfires lit for the ceremony of Las Posadas at Christmas in northern New Mexico, the fires at Chaco may have served as beacons to all travelers. The precipitous stairways leading down into the canyon seem to be more ceremonial or symbolic than practical, intended more for ritual and pageantry than transportation of food or building materials.

In the canyon some of the Great Houses may have provided water, wood, cooking utensils, cooked food, and shelter for pilgrims, and some may have been constructed to accommodate pilgrims from specific areas of the regional system and to provide secure storage for valuable ritual objects. Storage areas contained items for use in festivals such as the 110 cylinder jars found in room 28 of Pueblo Bonito.

Shelters for visitors to a pilgrimage center are ubiquitous features of pilgrimage traditions around the world. At the Kumbha Mela of Allahabad, thousands of temporary shelters (*dharmashalas*) are constructed and are generally designated for pilgrims from particular regions of India. More permanent *dharmashalas* for the use of pilgrims have been built in Varanasi by wealthy patrons. The size and style of these structures are meant to reflect favorably upon the piety and wealth of the donors and upon their places of origin.

Pilgrims and other visitors to Chaco Canyon may have traded with merchants who seasonally set up shop in the "road-related units" associated with a number of the Great Houses. At Pueblo Alto, near the terminus of the North Road, there are distinctive rooms that had been identified as "road-related units" that may be related to the activities of visitors. The units have very little variation between them and appear to have been constructed from a standard plan, as if there had been a well-established tradition of serving the needs of travelers. A large block of road-related units at the southern terminus of the North Road is found in the East Ruin, which is connected to Pueblo Alto by a large masonry wall.

Along the back wall of Pueblo Bonito, a row of some eighty rooms contains beams dated to 1040–1050 CE, which appear to be similar to some of the road-related units of Pueblo Alto. The rooms were attached to preexisting walls but did not have access to the rooms against which they were built. The possible uses of these road-related units include sleeping space for visitors; shops providing food, goods, and mementos to pilgrims; or storage facilities for food or other items associated with festivals.

In order for everyone to arrive at the same time, a system-wide calendar with an accuracy of a few days would have been needed. Such accuracy is not necessary for local events when word-of-mouth announcements would suffice. When travel of one to two weeks is involved, travelers need to know when to depart from their homes to avoid wasting time waiting at the destination or missing the festival entirely. Only a few such negative experiences would have been strong incentives for developing one's own calendar with appropriate accuracy. The requirements placed upon such a calendar are much more stringent than those of agriculture, in which yearly variations of several weeks in optimum planting dates may occur. Agriculturists, skilled in reading subtle changes in soil conditions and plant and insect life, do not need a highly accurate solar-lunar calendar.

Among the historical Puebloans, winter solstice was an important calendrical event because the people and the land depended upon the return of the sun to the north. Often the day of full moon closest to winter solstice was chosen for the major winter festival. Such an event may similarly have been a time for regional festivals in Chaco Canyon. That period of time was best for traveling because the increasing illumination of the waxing moon extends the day for nearly two weeks before full moon.

The calendar could have been established and corrected by direct observations of the horizon sun at a number of calendrical stations in Chaco Canyon, such as Piedra del Sol (summer solstice), Wijiji (winter solstice), and Kin Kletso (winter solstice). Line-of-sight communication from Chimney Rock via Huerfano Peak to Pueblo Alto could have provided further accurate dates, including those of the spring and fall equinoxes and summer solstice.

Violence, Warfare, and the Sun

Around 1250 CE, bad things began to happen among the canyons and mesas of the Southwest. There was a rapid movement of people living north of the San Juan River into defensible villages, often on canyon rims incorporating water supplies. Sometime around 1280 CE, some forty-one men, women, and children were violently killed at Castle Rock Pueblo, in a surprise attack apparently meant to annihilate the community.[6] Bodies were left unburied among the ruins of the village. The attackers may have been driven to violence by competition over diminishing resources. Castle Rock Pueblo had been built with defense in mind. The healed bones of some of the skeletons revealed that this final attack on the pueblo had not been the first. In Sand Canyon, 4.7 miles northeast of Castle Rock, there were eight violent deaths, with evidence of scalping, decapitation, excessive bone breaking, and unnecessary violence.

Recent studies by Timothy Kohler and Kathryn Kramer[7] on the gender differences of human remains suggests that some of the violence among the Pueblos in the eleventh to thirteenth centuries may have involved kidnapping women. Among the burials found in the Mesa Verde

FIGURE 2.7. Fajada Butte.

FIGURE 2.8. Hovenweep Square Tower.

region from the 1200s, there is a deficit of females, with twenty-two females and thirty-nine males. In the same period the sex ratios in certain sites of the San Juan basin were reversed, with an excess of women. If one aspect of warfare during this period across the Southwest was the capturing and abduction of women, one can view the towers of Hovenweep and the high ledges of Fajada Butte as places to hide women and children (Figure 2.7 and 2.8).

The majority of pottery among the thirty-five rooms on Fajada Butte is late 1200s Mesa Verde black-on-white;[8] these rooms appear to be part of a movement of people to "pinnacle" sites during this time.[9] People were seeking refuge elsewhere in the San Juan basin at other pinnacle sites such as Kin Nazhin, with fifty rooms, and El Castillejo, with forty-two rooms and nine kivas.

To the west in the Kayenta region, the Organ Rock Pueblo was built on top of a sandstone block with sheer walls 164 yards high, accessed only by a single crack, stairs, and handholds. There was no water or cultivable land on top. Jonathan Haas and Winifred Creamer comment that there could have been "no sense for people to live up there unless they were absolutely convinced of the need for protection." In their surveys of sites they found that "one of the best indicators of site location was when the survey crews were unable to find an access route to the top of a particular topographic feature."[10] These sites on the tops of isolated buttes invariably dated to 1250–1300 CE.

These locations could not have been successful sanctuaries during a protracted siege, since they lacked water and food. They would have provided short-term safe havens during raids. Fear of those raids must have been very powerful to force people onto the high ledges of Fajada Butte and other pinnacle sites. News of the massacre at Castle Rock around 1280 CE may have spread across the region. These are grim and unpleasant facts, but they are facts. They do shed light on the astronomy that we find among the towers of Hovenweep, the upper ledges of Fajada Butte, and defensive villages that were built in the latter half of the thirteenth century. The exquisite spiral at the three-slab site of Fajada Butte may have been produced at this time by people confined to its summit while seeking refuge from threats below and fervently hoping for assistance from the skies above.

CHAPTER 3

THE DOME
OF THE SKY

Our experience of time on this planet comes from the two major motions of the earth, one its spinning on its axis and the other its annual revolution around the sun. The spinning earth gives us clock time, sunrises and sunsets, and all those experiences of temperature changes and habits of sleep associated with night and day. The motion of the earth in its orbit around the sun provides us with calendrical time. It gives us birthdays, the seasons, the heat of summer, and the cold of winter. Driven by these two motions of the earth in space, the sun traces a daily as well as a seasonally changing path across the sky.

Events in the sky, whether by day or by night, appear to occur on the inside surface of a vast hemispherical dome. The early Greeks imagined the sky to be a series of spheres composed of indestructible crystalline material with stars embedded like small bright jewels. Using the wheel as a metaphor for the heavens, they imagined that the spheres rotated, with the earth at the unmoving center. The Greeks recognized that the sphere of stars had to lie at a great distance from the earth, because the stars did not change their positions as sky watchers moved across the surface of the earth.

Cultures that did not possess the wheel, such as those of Mesoamerica, divided the heavens above and the worlds below into parallel layers. The Pueblo cultures of the American Southwest have an intermediate cosmology consisting of layers of hemispheres. Beneath our world lie three or four dome-like worlds through which their ancestors passed in their struggle upward.

When we go out to look at the sky for a moment at night, we see only a half-sphere over our heads. But given time and patience enough, watching new stars rise in the east and old ones set in the west, we sense that a full sphere of stars surrounds us and that we lie at its center. Each of us seems to be at the center of a vast and slowly turning sphere of stars.

Unfamiliar and often confusing to us city dwellers, the paths of the stars and sun across the hemispherical sky must have been familiar and predictable to ancient sky watchers. Part of our amazement over the accomplishments of ancient astronomers results from our own night blindness, from our inability to see the skies in our brightly lit and building-enclosed world.

The sky is one of the major symbols in the natural world of order and cyclic repetition. Yet the translucent blue of the sky seems infinitely distant, and the gods living there seem infinitely inaccessible. In many cultures, people have transformed their homes and temples into miniature universes, which are smaller and more manageable than the larger reality. The circular kiva of the Pueblo peoples, the hogan of the Navajo, the dome of the Indian stupa and Tibetan chorten, and even the dome of St. Peter's mimic the celestial sphere.

The Kiva as an Astronomical Symbol

Ever mysterious in her cycles of life and death, in her power to provide life and then take it away, Mother Nature is an inscrutable benefactress, not always benign or fully predictable. To the Ancestral Puebloans living along the northern frontier, the world must have often been threatening and dangerous. Especially during periods of climatic instability, agriculture would have been unpredictable and frustrating. Sometimes the growing season may have been so short that the crops yielded little food. The animals may have been strangely absent. The anticipated rains may never have arrived. Overlying this uncertain life, moving smoothly and confidently across the sky, the regularity of the sun was in clear conflict with the uncertain chaos of the land beneath.

It was such a distinction between the apparently unchanging order of the heavens and the painful change and decay of the earth that led Plato and Aristotle to separate the two realms. The four elements of the Greeks—air, fire, earth, and water—moved with straight-line motion on the earth; the fifth crystalline element, ether, moved in endless circular motion in the heavens. The Ancestral Puebloan may have adopted another strategy to deal with the conflict between heaven and earth, building circular kivas as copies of the heavens. With a dome overhead aligned to the four cardinal points of space, the microcosm of the kiva may have been a place to achieve harmony with the larger world.

FIGURE 3.1A. Great Kiva of Casa Rinconada. The typical great kiva contained a raised fire box, raised oblong floor vaults placed between the supports for the roof columns, and a north ground-level room reached via a staircase.

FIGURE 3.1B. Kiva at Yellow Jacket, 5MT-2. The characteristic features of kivas of the northern Ancestral Puebloans are a sipapu, hearth, deflector stone, southern recess, and six pilasters along the perimeter.

The kiva is the most characteristic feature of Pueblo and Ancestral Puebloan architecture (see Figure 3.1). Like the sacred structures of India and Mexico, as well as others around the world, it provides both a cosmology and a cosmogony, a description of the world as it is and a theory of how the world came into being. According to modern Pueblo tradition, the kiva represents the *sipapu,* the place where the first humans emerged from the lower worlds.[1]

> *When they built the kiva, they first put up beams of four dif-ferent trees. These were the trees that were planted in the under-world for the people to climb up on. In the north, under the foundation they placed yellow turquoise; in the west, blue turquoise, in the south, red, and in the east, white turquoise. Prayer sticks are placed at each place so the foundation will be strong and will never give way. The walls represent the sky; the beams of the roof (made of wood of the first four trees) represent the Milky Way. The sky looks like a circle, hence the round shape of the kiva.*[2]

Itself a symbolic point of emergence of the ancestors, the kiva may also contain a sipapu in its floor, slightly north of center. The four pillars of the Ancestral Puebloan great kiva may have represented the trees

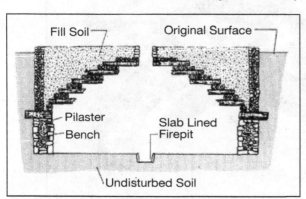

FIGURE 3.2. Cribbed roof of a kiva, side view.

planted in the under-world or the four sacred mountains. The kiva thereby may have been a link between heaven and earth as well as a link between past and present.[3]

In Chaco Canyon, many kivas in Pueblo Bonito had cribbed roofs made of horizontal logs placed at regular intervals along low benches (Figures 3.2 and 3.3). As many as 350 pine logs were used in the kiva roofs, most if not all of which had been carried or dragged over large distances.[4] Kivas range in diameter from 10 feet to over 60 feet, as is the case of the great kivas at Yellow Jacket and Casa Rinconada in Chaco Canyon.[5] The amount of labor needed to construct these primarily subterranean great kivas was enormous and indicates the great importance the Ancestral Puebloans gave to the construction of ceremonial buildings. In the great kiva of Casa Rinconada, the four pillars supporting the roof were carefully aligned to the four cardinal directions (Figure 3.4).[6] Precision of orientation appeared necessary to achieve a suitable copy of the sky.

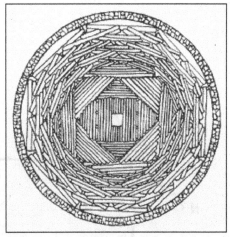

FIGURE 3.3. Cribbed roof of a kiva from inside looking up.

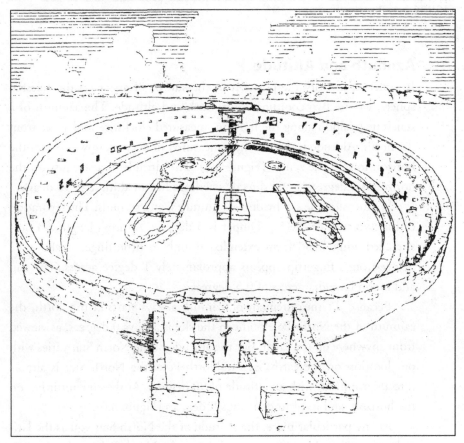

FIGURE 3.4. Great Kiva of Casa Rinconada.

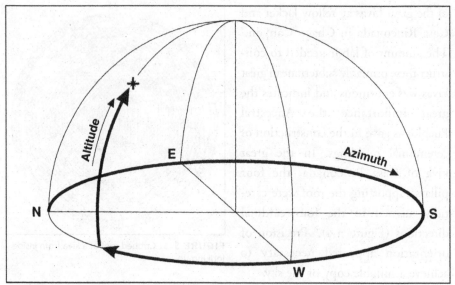

FIGURE 3.5. The celestial sphere.

Azimuth and Altitude

In order to identify the position of the sun or of a star on the celestial sphere, we use two coordinates, azimuth and altitude. The azimuth of a star is its position along the horizon, measured clockwise in degrees from the north. Altitude is the distance, measured in degrees, upward from the horizon toward the zenith (Figure 3.5). The altitude of a star just on the horizon is 0 degrees; a star directly overhead has an altitude of 90 degrees.

We measure the sky by degrees, minutes, and seconds. The separation of the two stars of the Big Dipper is 5 degrees. At arm's length, an outstretched adult hand from extended thumb to little finger is about 18 degrees; one's fingertip appears approximately 1 degree across. The sun and moon have diameters of 0.5 degrees.

Because the measurement of azimuth always starts at the north, the azimuth of the North Pole point in the sky is always 0 degrees, as viewed from anywhere on the earth. But the altitude of the North Star varies with our location on the earth. At the North Pole, the North Star is almost directly overhead, with an altitude of 90 degrees. At the equator it lies on the horizon, and its altitude is thus 0 degrees (Figure 3.6).

At any particular place, the altitude of the North Star equals the latitude of that location. As one travels from the North Pole to the equator,

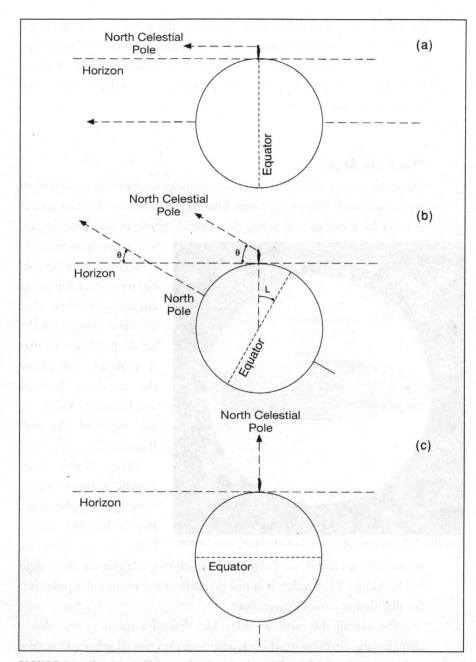

FIGURE 3.6. The altitude of the north celestial pole at different latitudes.

the North Star thus falls in the sky from the zenith to the northern horizon. At the equator of the earth, which has a latitude of 0 degrees, the altitude of the North Star is also 0 degrees. At Yellow Jacket, with a latitude of 37.5 degrees, the pole of the heavens lies 37.5 degrees above the northern horizon.

The Pole Star

The earth rotates on its axis in a counterclockwise direction as seen from above the North Pole of the earth. Due to that rotation, the stars appear to move from east to west across the celestial sphere. In one hour the sun

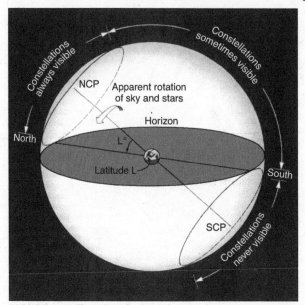

will move approximately 15 degrees. There is a special set of stars, known as circumpolar stars, that are close enough to the north polar point that they never sink below the northern horizon and hence are visible on any night of the year (Figure 3.7).

The North Star, Polaris, is today one of these circumpolar stars. But it has not always

FIGURE 3.7. The celestial sphere showing the circumpolar stars.

been so, nor have we humans always had a North Star with which to guide our travels or align our buildings. Even today, it is not precisely at the north polar point but lies slightly less than a degree away.

The axis of the earth wobbles, like that of a spinning top, slowly changing the direction to which it points on the celestial sphere. That slow wobble, which takes 26,000 years to complete a full cycle, is known as precession (Figure 3.8). As a result of the wobbling, the north polar point slowly sweeps through the sky. During most of the past, there has been no bright star near the north polar point, just as there is no bright star currently near the south polar point. As we look backward in time, the distance of Polaris from the north polar point was greater. At 1200 CE, Polaris was 5 degrees away from the pole; at 900 CE it was 6.7 degrees away, and at 0 CE, it was nearly 14 degrees from the pole (Figure 3.9).

The Ancestral Puebloans organized their buildings and living spaces on a north-south line without the assistance of a bright North Star. Other methods, more sophisticated than simply taking a sight upon a star, would have been needed. Observing the position of sunrise and sunset on a given day would not have worked unless the horizon was perfectly flat.

FIGURE 3.8A. Precession of a top.
FIGURE 3.8B. Precession of the earth.

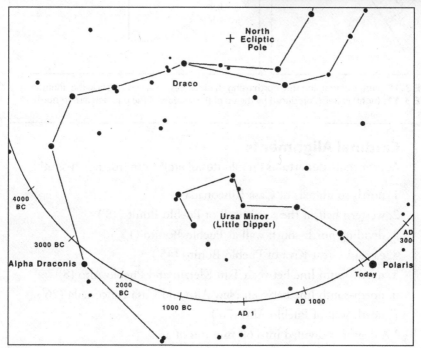

FIGURE 3.9. Movement of the celestial pole due to procession.

A pole, called a gnomon, may have been placed vertically in the ground by the Ancestral Puebloan astronomer-priest and used to determine geographic north (Figure 3.10). The gnomon, defined as a vertical, shadow-throwing pole, was used by many cultures—Borneo tribespeople, Babylonians, Greeks, and Chinese—to establish the length of the year and the time of the solstice by measuring the length of the shadow. The Kogi, living in the foothills of the Colombian Andes, built temples with doors opening east and west using the gnomon. Likewise, the architects of the great temples of India used shadows cast by upright posts to orient their structures to the cardinal directions.

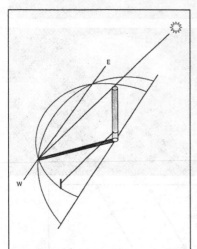
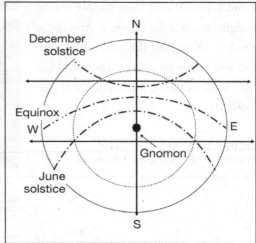

FIGURE 3.10. (left) Determining true north using shadows cast by a vertical stick, the gnomon. **FIGURE 3.11.** (right) Patterns produced by the tip of the shadow of the gnomon at mid-northern latitudes.

Cardinal Alignments

Approximate departures (in minutes of arc)* from true north-south:

1. north-south axis of Casa Rinconada (4')
2. western half of the south wall of Pueblo Bonito (8')
3. dividing north-south wall at Pueblo Bonito (15")
4. eastern Great Kiva of Pueblo Bonito (45')
5. north-south line between Tsin Kletzin and Pueblo Alto (8')
6. north-south line between New Alto and Casa Rinconada (26').
7. north wall of Pueblo Alto (70')

* A degree is divided into 60 minutes of arc.

The gnomon may have been used by the Ancestral Pueblo peoples in the following manner: Throughout the day, the end points of the pole's shadow could have been marked by small stones or sticks. A rope attached to the pole could have been used to mark out a circle that cuts across the stones. A line drawn to connect the two places cut by the circle would run east-west; the line perpendicular to that direction would be north-south. If all these steps were performed carefully, if the ground on which the shadows were cast was carefully leveled and the stick had been aligned accurately to the vertical, then the method should have allowed the Ancestral Puebloans to align their structures with the kind of accuracy we have discovered among their ruins (Figure 3.11).

The Celestial Equator

Like the earth, the sky has an equator, the celestial equator, which is the outward projection of the earth's equator. From the earth's equator, the celestial equator is directly overhead. Any object that lies on the celestial equator rises precisely in the east and sets precisely in the west. At the time of equinox, the sun is on the celestial equator and spends twelve hours above the horizon.

On the earth we determine how far north or south of the equator we are by latitude: northern or positive latitude for the Northern Hemisphere and southern or negative latitude for the Southern Hemisphere. The distance of the sun, a planet, or a star from the celestial equator is known as its declination, an exact counterpart to latitude on the earth. At the time of the equinoxes, March 21 and September 21, the sun is exactly on the celestial equator, and its declination is 0 Degrees. At the time of summer

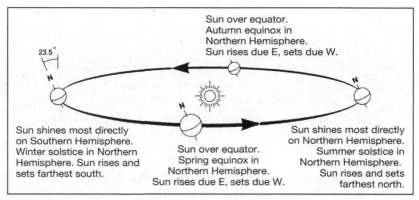

FIGURE 3.12. The seasons of the earth.

solstice in the northern hemisphere, the sun is farthest above the celestial equator and has a declination of +23.50 degrees; at the winter solstice, its declination is −23.50 degrees (Figure 3.12).

The Path of the Sun

Even though the movements of the stars as they rise in the east and set in the west are very regular, the sun, the moon, and the planets wander across the celestial sphere in less regular patterns. Throughout the year the sun moves from west to east along its own particular pathway known as the ecliptic. The ecliptic is tilted by 23.50 degrees from the celestial equator, and the sun takes one year, or 365.24 days, for a complete circuit of the ecliptic.

Relative to the background stars, the sun moves about 1 degree per day, which amounts to twice its diameter per day. As a consequence, constellations shift their locations relative to the sun throughout the year. Each night a given star rises a few minutes earlier; at sunset, new constellations appear on the eastern horizon and move up higher and higher each night.

The Rhythmic Sun

At the times of the fall and spring equinoxes the sun spends equal amounts of time above and below the horizon. On those two days it moves across

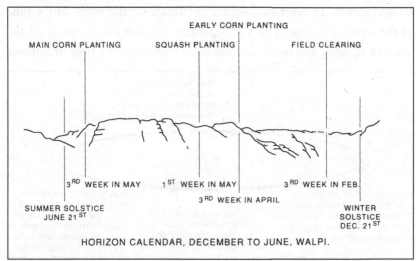

FIGURE 3.13. The Hopi solar calendar.

the sky on the celestial equator, rising in the east and setting in the west. Only if the horizon is perfectly flat will the rising and setting positions be exactly east and west. On the two days of equinox, the sky is divided into four parts: east and west, which are approximately established by sunrise and sunset; north and south, established when the sun is highest in the sky. Such a division of the dome of the sky into the four quarters clearly had great symbolic significance for ancient sky watchers, as judged by the number of peoples who aligned themselves and their buildings along the four cardinal directions. Many cultures, such as the ancient culture of India described in the Vedas, considered March 21 to be the start of the new year. Equinox was the time for the Vedic priest to rebuild a ceremonial altar made of 360 bricks in order to rejuvenate the old, exhausted year.

During the spring, the sun rises each day farther north along the horizon and at noon reaches higher in the sky. Once the extreme has been reached, at the time of the summer solstice, the sun pauses at its most northerly rising position for several days and then begins its downward motion toward winter solstice. As the days grow shorter, the sun's declination decreases until December 21.

This oscillation of the sun between summer and winter sunrise points along the horizon gave ancient astronomers a convenient calendar (Figure 3.13). For those who could read it, the horizon was a reliable timepiece. If the horizon was distant and contained sharp peaks and clefts, the day of the year could be accurately determined at sunrise or sunset.

The cyclic motion of the sun is variable and repetitive: fast at equinox and slow at solstice. Reflected in the seasons—with the changes in rainfall, duration of the day, temperature, and growing season—the sunrise positions were intimately linked with human life and the environment. How could they be ignored?

The sun does not move at a uniform rate along the horizon but noticeably slows when it approaches solstice, like a pendulum slowing at the end of its swing. At the latitude of Yellow Jacket, at the time of equinox, the sun moves a full solar diameter, half of a degree, between successive days. Near solstice the sun's motion drops to zero, returning to practically the same sunrise point each day for a week. Solstice means "stand still" in Latin.

There was a potential danger in the stopping of the sun at the two end points of its swing along the horizon. At the time of winter solstice, among the historical Pueblo peoples there was concern that the sun would cease its motion forever, or even worse, that the sun might fall off the edge of the world. The land might sink into a cold of unending winter. Festivals were organized to send the sun on its way.[7] At summer solstice, there was concern that the sun would pause too long at its summer house, throwing off the proper cycle of the year, shortening the agricultural year, and permitting a freeze before the time of harvest.[8]

The Rhythmic Moon

In any given month, the rising moon swings between two extremes on the eastern horizon, similar to the oscillation of the rising sun during one year. The orbit of the moon is tilted with respect to the earth's orbit by 5 degrees, 9 minutes; perturbations cause it to vary from 4 degrees, 57 minutes, to 5 degrees, 20 minutes. As a result, the moon can rise farther to the north of sunrise on summer solstice. Whereas today the sun reaches a maximum declination of 23 degrees, 26 minutes, the moon can reach a declination of 28 degrees, 46 minutes. When the moon reaches its maximum northern or southern declination, it has a "standstill" similar to that of the sun near the solstice. Due to the gravitational effect of the sun, the moon's orbit wobbles over a period of 18.61 years, like a juggler's spinning plate suspended by a pole.

The major standstill is followed by the minor standstill 9.3 years later, when the moon swings in one month between declinations of ±18 degrees, 7 minutes (Figure 3.14). As astronomical phenomena go, minor standstills are rather underwhelming and difficult to detect. During major standstills, the moon can rise and set at places where the sun can never reach, whereas at minor standstills, the moon is doing nothing unusual, rising at places the sun reaches two times a year.

Major standstills of the moon may have been detected by prehistoric inhabitants of northern Europe, as suggested at Stonehenge and other megalithic observatories, although the evidence remains ambiguous and, indeed, has been vigorously contested by some critics.[9] At high latitudes the swing of the moon is much amplified, such that there is a lunar version of the arctic circle, inside of which the extreme southern standstill moon does not rise. In the recumbent stone circles of Scotland, the southern standstill moon would have just slid along the top of a sacrificial stone.

The first suggestion that the Ancestral Puebloans were aware of major or minor lunar standstills came from studies of the three-slab site on Fajada

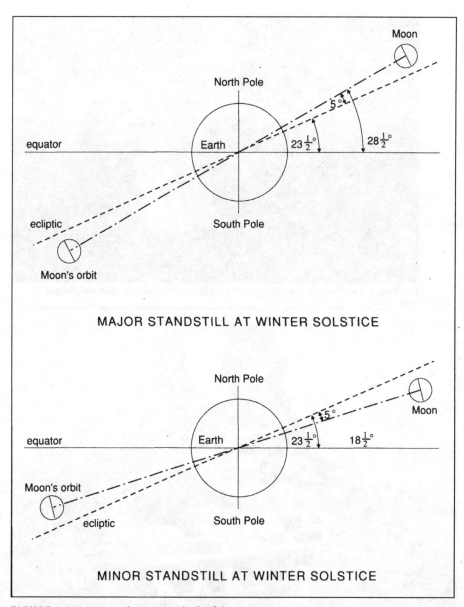

FIGURE 3.14. Major and minor standstills of the moon.

Butte.[10] These suggestions were received with little enthusiasm by the archaeoastronomical community.[11] As I shall discuss in Chapter 5, there are a number of reasons for skepticism that lunar standstills are marked on Fajada Butte. However, Ancestral Puebloans may have discovered lunar standstills at Chimney Rock in the latter half of the eleventh century, which is more thoroughly discussed in Chapter 6. Here it would be helpful to point

FIGURE 3.15. The major standstill moon rising between the double chimneys, August 8, 1988.

FIGURE 3.16. Chimney Rock great house.

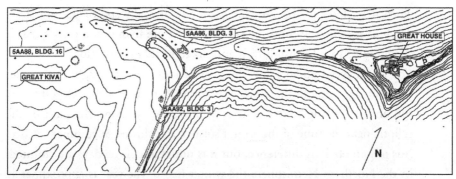

FIGURE 3.17. Map showing the relationship of East Kiva and great house at the Chimney Rock mesa. Note that the great house is separated from the residential area to the west.

out that Chacoans built the most remote great house of all their outliers on the upper mesa of Chimney Rock and that it had an excellent view of the major standstill moon rising between the double chimneys (Figure 3.15).

At the time of a major lunar standstill, the moon's swing is greater that that of the sun, rising to the north of the sun at June solstice and to the south of the sun at December solstice. A distant, irregular horizon is very helpful in discovering this effect. It would be difficult to see the difference between the sun and moon on a perfectly smooth horizon, such as is present in Chaco Canyon or above the jungles of the Yucatan. The horizon at Chimney Rock is ideal, for at its extreme the moon rises between double towers, something that the sun can never accomplish.

This visually dramatic event could have first been viewed by early residents on the high mesa of Chimney Rock between 1055 CE and 1057 CE. Perhaps the residents of the great houses in Chaco were informed about these spectacular moonrises by traders moving between Chimney Rock and Chaco. During the next standstill cycle, in 1076 CE, a team of Chacoans built an isolated kiva (now the East Kiva) on the high mesa at a location where the moon could be seen rising between the towers (Figure 3.16). Then, at the next standstill cycle, in 1093 CE, the Chimney Rock great house was constructed, signaling the full incorporation of Chimney Rock and lunar standstills into the Chacoan cosmos (Figure 3.17).

Sun Time and Clock Time

The earth moves on an elliptical, not a circular orbit, around the sun. As a result, the speed of the earth in its orbit is not constant. Moving faster when it is closer to the sun, the earth rhythmically speeds up and slows down as it swings around the sun. Since we are living on a planet that is changing its speed constantly, it appears to us that the motions of the outside world speed up and slow down at various times throughout the year.

During winter in the Northern Hemisphere, the earth is closest to the sun and is moving fastest. The sun moves fastest along the path of the ecliptic near the time of the winter solstice and slowest at summer solstice. It is not an obvious difference, but it is noticeable if the days are counted. In the Northern Hemisphere the winter half of the year is approximately eight days shorter than the summer half.

Because of this variation in the speed of the earth and the apparent speed of the sun in the sky, the time determined by a sundial usually differs from that of a wristwatch, which runs at a constant rate. The wristwatch runs on a fictitious sun, called the mean sun. When the mean sun is due south and highest in the sky, it is called noon, local mean time.

Ancient people would have had a sense of the cyclic nature of the seasons, and they probably would have experienced time as a variable feature of their lives. We have made time uniform and rigid through our technology of the clock, preferring the convenience of gears and crystals to the varying shadows of the sundial. As we shall see, the Ancestral Puebloans constructed their own unique versions of sundials, using the varying play of light and shadow upon stone to establish and celebrate the changing position of the sun on the dome of the sky.

CHAPTER 4

SKY WATCHERS

For clues to the techniques and meaning of Ancestral Puebloan astronomy, we turn to today's Pueblo Indians. It is reasonable to assume that some astronomical traditions have been carried on since Ancestral Puebloan times, although perhaps others have been lost. The modern Pueblos are descended from a mix of prehistoric southwestern peoples and doubtless have absorbed traditions from all of them. They have also experienced pressures, usually less than friendly, from nomadic Navajos, catholicizing Spaniards, and land-hungry Anglos that have resulted in cultural adaptations that the Ancestral Puebloans could not have imagined.

Some of the differences between the Ancestral Puebloan and the modern Puebloans are obvious. Great kivas have disappeared from Puebloan architecture, while the elaborate masonry of the Ancestral Puebloans has been largely supplanted by adobe bricks. The roadways and other features indicative of a centralized Ancestral Puebloan authority have also disappeared. Whatever traditions the modern Puebloans may have retained of their ancient ancestors—the culture responsible for the Mesa Verde cliff dwellings, the Hovenweep towers, and the Chaco great houses—have been transmuted into that rich mixture of past and present that are today's Pueblo societies.

Today, no two Pueblos are alike, and no single voice can speak for the past. Pueblo peoples speak variations of four distinct languages, and the largest linguistic group, Tanoan, has three subgroups. Western Pueblos, such as the Hopi and Zuni, tend to be organized by kinship groups, whereas eastern ones, such as the Tewa, are dual in structure, with

"Winter" and "Summer" people alternately dominating religious affairs. Western Pueblos emphasize solstice celebrations more than eastern ones. And although some groups have circular kivas similar to those dominant among the Ancestral Pueblo, others have rectangular ones. Still others have none at all.

The upshot of all this is that even though the nature of modern Puebloan society can provide insights into the lives of the Ancestral Puebloans, analogies between historical and prehistoric practices should be drawn only with care. For example, anthropological reports indicating that Hopi, Zuni, and Keresan priests used light and shadow effects to date solstices increase the likelihood that such effects at Ancestral Pueblo sites were intended for similar purposes. Observations that contemporary Puebloan peoples do not notice lunar standstills or record unique astronomical phenomena can weaken theories about the use of the Chacoan "sun dagger" or the supernova content of the Penasco Blanco painting. However, because many Ancestral Puebloan traditions may have been lost or changed over the centuries, such observations do not definitively rule out the possibility of these ancient practices.

Sun Watching

Sun watching is one of the few traditions that can safely be termed pan-Puebloan.[1] Historical Pueblo Indians have watched the movements of the sun along the horizon, have observed the play of light through windows and portholes at crucial times of the year, and have established sun shrines at key locations. Although observation stations are usually located in or near pueblos and are often unmarked, shrines are less accessible and frequently resemble cairns or other man-made structures. A religious official is generally responsible for making the anticipatory and confirmatory observations for important dates. Zuni Pueblo used to have a special Priest of the Sun, called the *pekwin,* but when the last one died, his duties were assumed by the Rain Priest. At the Hopi villages, the head of the society responsible for the upcoming ceremony makes the solar observations necessary to determine the ritual's date. In the eastern pueblos, the Winter and Summer chiefs watch the sun during their respective periods of office. Finally, the War Chief is responsible for sun watching at a few Keresan pueblos.

Early anthropologist Frank Hamilton Cushing reported that the Zuni *pekwin* made a daily pilgrimage to a nearby Zuni ruin to watch the sunrise as the spring equinox approached. The *pekwin* sat in a ruined tower near a pillar bearing sun, moon, and star markings similar to those near Penasco Blanco, and a companion kept a count of the days remaining until equinox by notching a stick. Other methods of anticipating a crucial date by counting the days from a sunrise over a special feature of the horizon have included untying knots in a string and marking beams or wall plaster. The Zuni priest confirmed the equinox by noting when the shadows of what Cushing called "the solar monolith," the nearby Thunder Mountain, and "the pillar of the gardens of Zuni" were aligned. Meanwhile, Cushing reported that residents of Zuni were checking up on the Priest's accuracy by observing the play of light along plates imbedded in the walls of their homes. A window or porthole similar to those at Hovenweep Castle or possibly at Pueblo Bonito allowed the rising sun to illuminate a certain spot only on the crucial dates.[2] Some of the Hopi villages have used similar devices for predicting dates.

More commonly, sun watchers have simply observed the sun's movement along the horizon at either sunset or sunrise from any convenient spot. At one Rio Grande pueblo, the area in front of the Catholic Church was a sun station; at Hopi, the roofs of clan buildings often serve.[3] Knowledge of where to stand is apparently the private property of the observer and is not necessarily marked for posterity by glyphs or architectural features. Similarly, the decision of when to begin the observations seems, at least at Hopi, to be up to the discretion of the watcher. One sun watcher reported that he began his observations from the time when the sun set behind a motel on Second Mesa![4]

Sunrise is usually the crucial time of day for horizon observations, as well as light and shadow observations, but practice varies according to the pueblo. At the Hopi villages, the eastern horizon is generally watched for ceremonies falling after winter and up to and including summer solstice. But for ceremonies falling after summer and including winter solstice, the Hopi watch the western horizon. Part of the reason for this difference may be that the San Francisco Peaks to the southwest of the Hopi villages provide a distinct horizon calendar. Furthermore, the Hopi kachinas, or masked deities impersonated by Hopi men, arrive at the villages near the winter solstice and are said to reside in the San Francisco Peaks. The

Zuni apparently watch the sun at both sunrise and sunset, and the eastern Puebloans primarily watch at sunrise, although some of the Tanoans watch at sunrise for one half of the year and at sunset for the other. In addition to the sun-watching stations, most Puebloans place shrines at key points along the horizon.[5] Cushing considered the tower at the Zuni ruin to be such a shrine, but because it was located near a former pueblo, it was probably a typical sun station. Shrines are often located out of sight from the pueblo and require some exertion on the part of the priest or initiate when he goes to deposit prayer offerings at them. The western Pueblos seem to locate these sacred spots in the solstitial directions, although any site in line with an important horizon marker may qualify as a shrine. At Hopi and Zuni, the solstitial directions are seen as the cardinal ones, whereas the equinoctial directions are more important along the Rio Grande.[6]

Moon Watching

Among all of the Pueblos, the moon figures importantly in both cosmology and mythology.[7] Cloaked variously in male and female personas, the moon is an intermediary between the sun and the earth, able to exert a favorable influence on the giver of life.[8] Thus, the historical Puebloans have been glad to see the moon in the daytime sky, whereas eclipses, when the moon would hide his or her face from the people, were frightening experiences. The historical Puebloans evidently have been unable to predict such terrible occurrences.[9] Whether it is viewed as a male or female entity, the moon is generally associated with fertility in humans, animals, and plants. Much of the Puebloan moon lore revolves around women and children, and pregnant women are most vulnerable to the dangers of an eclipse. Near winter solstice at Zuni, prayer sticks, or symbolic offerings, are planted in the fields for both the sun and the moon, with women directing their offerings at the guardian of the night sky.

At Zuni, the *pekwin*'s job was greatly complicated by the necessity of holding the winter solstice ceremony when the moon was full. The Zuni felt it important that the moon at its strongest should buttress the sun at its weakest; during summer solstice, the opposite held true, and the "strong" sun complemented a "weak" crescent moon. This emphasis on balance often meant that the dates for the solstitial celebrations did not correspond exactly to the actual solstices. Similarly, the important solstice-related celebration of *Shalako* was ultimately dated by solar observation, but preliminary rituals were performed according to the phases of the moon for nearly two months beforehand.[10] Among the Hopi, each important annual ritual, although timed by the sun, was assigned a specific month. Months bore names indicative of the rituals, activities, or natural phenomena unique to them.[11]

Eastern Puebloans watched the moon carefully for weather omens, with variously shaded rings corresponding to different gradations of cold weather. Like the Zuni, the Tewa, a Tanoan subgroup, have noted that the moon's path is similar to that of the sun, but it is only during full phases that it is seen to mimic the sun precisely. Seasonal names dominate for eastern Puebloan months, where appellations incorporating agricultural elements or weather patterns are common. Among the Zuni and the Hopi, the lunar element of timekeeping is more prevalent, but this is probably because these pueblos were spared the overwhelming contact with the Spaniards that the Rio Grande groups experienced. The Hopi may once have even timed certain rituals by a lunar horizon calendar.[12]

Although accounts are varied, most Puebloans seem originally to have recognized thirteen months as composing a year. More manageable than the seasons and more practical than the days, the months were convenient timekeeping units. All pueblos seem to have conceived of the first visible crescent as the beginning of a particular moon; the days of the invisible new moon were left uncounted. The problem of reconciling the lunar year with the solar was usually resolved by leaving one or two months unnamed and simply ignoring one of them if a discrepancy arose. Sometimes an additional "short" month would be incorporated to resolve the difference.

The Hopi and the Zuni had five and six named months, respectively, followed by five or six more bearing the same names as the first series. For the Hopi, this practice represented the duality that underlaid the structure of the universe. While the winter fields lay fallow, residents of the underworld were said to be cultivating and harvesting crops. The harvest obtained in the underworld, good or bad, would be duplicated in this world during the fall months. Similarly, while the Hopi were celebrating the summer solstice with a relatively meager ceremony, the elaborate winter solstice rituals were underway in the underworld, and while *Soyal*, the Hopi winter solstice festival, was in full swing, denizens of the underworld were holding the celebration of summer. Thus, the upper and lower worlds were mirror images of each other.[13]

Stargazing

Evidence of prehistoric Puebloan star watching is scanty at best, although many rock art panels do have stellar themes. Unlike sun watching, stargazing does not require that the observer stand in a certain place, nor is the precision of light and shadow techniques necessary for the use of a star calendar. Among contemporary Puebloans, however, evidence for an avid interest in the night sky abounds.[14] Like those of many peoples, the cosmology of the Puebloans is headed by a supreme Sky God, who is omnipotent and all-encompassing. It was he, or he/she, as the deity is sometimes perceived, who initiated creation. But this god has retreated to the farthest reaches of the heavens, become abstract and inaccessible. The lesser gods, such as the sun, continue the work of the creator, interacting with humans and fueling the mechanisms of life.

Because it appears sometimes at dawn and sometimes at dusk, Venus often represents the twin gods of both Pueblo and Navajo traditions. Navajo myth describes the Holy Twins, Slayer of Alien Gods, and the Child Born of Water. These twins were sons of the sun, and their exploits resulted in the establishment of many of the rules by which humans still live. Thus, the great star of morning and evening, Venus, was the patron of warriors and of hunters; today, Pueblo peoples send it prayers for the fertility of their farm animals. Finally, the morning star, in conjunction with a crescent moon, has appeared on the masks of important Hopi kachina dancers, symbolizing the solstice.

The Rio Grande Pueblos watched the paths of the Great Bear, Orion, and the Pleiades, relying on the regularity of their movements for ordering their nocturnal rituals. In some cases, these rituals were timed by the passage of these constellations over the kiva hatchways. Tewa religious officials anxiously watched for the rising of Orion's belt near both solstices, believing that an early May appearance would mean a long growing season. Seen as a bridge between this world and the celestial realm, the Milky Way remains important to all of the Pueblo peoples and may once have been viewed as a god in its own right. Possibly, it, too, was once a metaphor for the omnipresent Sky God.[15]

CHAPTER 5

CHACO CANYON

Within its walls, Chaco Canyon contains the full sweep of the astronomy of the Ancestral Puebloans, from the alignment of buildings to the north, to attention to solstices, and, finally, to the building of a refuge high on Fajada Butte during the dangerous days of the late 1200s.

At first, Chaco Canyon may seem a dry and inhospitable wilderness. But it should take only a dramatic sunset or a day of wandering beneath the cliffs among shifting hues in sky, sand, and sage to realize that this empty desert is a place of quiet majesty and beauty. About 1,000 years ago,

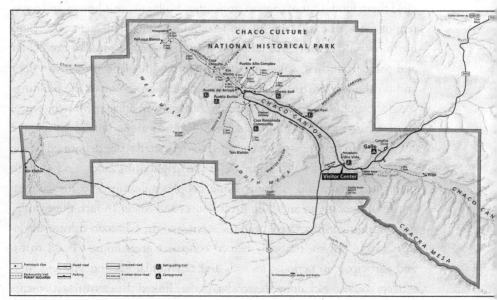

FIGURE 5.1. Chaco Canyon.

The Astronomy of Chaco Canyon
Places to Visit

Solstice sunrise: Casa Rinconada
                    Piedra del Sol

Any day: Horizon marker at Wijiji
                    Three faces of Piedra del Sol
                    Petroglyph below Peñasco Blanco
                        (supernova and/or Halley's Comet)
                    Horizon marker at Kin Kletzin
                    Pueblo Alto: bedrock basins along the
                        trail; east-west wall of the great
                        house; views of roads; view of
                            Huerfano Peak
                    Tsin Kletzin: north-south alignment
                        to Pueblo Alto

people turned the area into the center of a vast political and religious system that extended 250 miles from south to north and 190 miles from east to west. They guided scarce rainwater to their crops and made the earth yield them food. Out of canyon stone, they built tall great houses and a system of roads that still defy full understanding.[1] The settlements are crumbled now, and centuries of rain and snow have erased most of the plaster and paint from their walls, but the graceful buildings on the canyon floor and on the canyon rims tantalize the imagination. Some of the buildings recall the architecture of the Aztecs and Mayans of ancient Mexico; others are reminiscent of Tibet and Ladakh. Still others are unique to Chaco.

Chacoan culture was not confined to the sacred space delineated by the canyon's variegated walls. Throughout the neighboring lands, extending to Colorado, Utah, and Arizona, some peoples subscribed to the cultural norms of the canyon and built great houses and roads. In fact, these outlying communities may have been the essence of the Chacoan culture, as visitors to the canyon probably helped construct great houses, build roads, and participate in periodic festivals and ceremonial feasting. Ancient messengers could travel along the roadways that linked the canyon settlements with the outlying ones.[2] Traders and pilgrims could have walked into the canyon to participate in periodic trade fairs and festivals. Thirty feet wide in places, edged with low walls, these roads ran in remarkably straight lines radiating outward from Chaco Canyon, attesting both to the engineering skill and the self-image of the society that built them. As the Ancestral Puebloans never invented the wheel and had no beasts of burden, the purpose of these roads remains enmeshed in the many mysteries of Chaco.

Chacoan communities were linked by more than roads. Bonfires may have flared from ingeniously situated towers and shrines, reminders to important towns and villages of their ties to the center, perhaps communicating important dates (such as the full moon near the winter solstice), and announcing imminent festivals in the canyon. Whatever the reasons for the elaborate buildings, the intricate webbing of roads, and the signal shrines, they could only have been constructed by a highly organized and integrated society.

We have ample evidence of the craftsmanship and skills of the ancient Chacoans, but what was the source of their influence and power? What was the source of the charisma of their leaders to have such influence? Who, indeed, were their leaders? Why was it so important to construct their great houses with such care, aligning them to the larger cosmos? Were they primarily clever architects who designed buildings that required little maintenance, smart traders who carried their profession to unprecedented heights in the Southwest, or were they hard at work seeking meaning in their cosmos? What indeed would our world look like through the eyes of a living Chacoan?

Of one thing Chacoan we can be certain: the sky was of special importance. Mysterious powers and animate beings resided in the heavens. From the sky came rainwater and sunlight, both of which were essential to survival in an agricultural society. Chacoans needed to know when to prepare the irrigation systems for the onslaught of the summer rains and when to steel themselves for the winter months. They needed to know how to orient some of their important buildings along the north-south line, so as not to live against the grain of the cosmos. These needs, together with the sheer pageantry of sunrise and sunset in the desert, could have made sun watching one of the central activities of the priestly Ancestral Puebloan astronomers. The practicality of observing the sun's annual celestial journey, so crucial to survival in the marginal desert climate, was undoubtedly inseparable from the religious significance of those observations. In a desert world where life must be coaxed out of the environment, the sun's "decision" to return from its winter home, as well as its actual arrival at the northern extreme of its journey, is an occasion for celebration and thanksgiving.

The Spectacular Skies of Chaco

The first great houses appeared in Chaco in the late 800s, and their residents could not have missed seeing the most brilliant stellar outburst ever recorded by humankind. The supernova that first appeared on May 6, 1006, was reported in China, Japan, Korea, the Arab lands, and Europe.[3] This supernova reached a maximum magnitude of approximately –9.5 during the three months when it was brightest. Astrologers interpreted it as a harbinger of warfare and famine. The third brightest object in the sky next to the sun and full moon, this supernova ranks as the greatest astronomical event in human history. Soon after it first appeared, it could be seen during the day, and it was bright enough at night to read by and cast shadows. It continued to be visible for nearly two years.

Astronomical magnitude is a peculiar and counterintuitive brightness scale that has been used by astronomers for more than two thousand years, and deserves an explanation at this point. Around 129 BCE, the Greek astronomer Hipparchus classified stars on a numerical scale such that the brightest in the sky were first magnitude; stars not quite so bright were second magnitude, and the faintest he could see were sixth magnitude. The scale is actually based on the nature of human sense perception, which has a logarithmic response to stimuli. A doubling in intensity of a response results from more than a doubling in the intensity of the stimulus. For the human eye, for example, a second magnitude star is actually 2.5 times fainter than a first magnitude star, and a sixth magnitude star (the faintest that can be seen by the unaided human eye) is 100 times fainter than a first magnitude one. One can perceive fainter stars by using bigger eyes. For example, typical 50-mm (nearly 2-inch) binoculars will enable one to see ninth magnitude stars. A telescope with a diameter of 6 inches will reach to the thirteenth magnitude. The Hubble Space Telescope with a diameter of 94 inches can capture images of distant galaxies and quasars that are as faint as +31st magnitude! Very bright planets and stars have *negative* magnitudes and the very brightest natural object in our heaven, the sun, has a magnitude of –26.7.

The 1006 CE supernova was far south in the sky with a declination of approximately −37.5 degrees, in the southern constellation of Lupus. At the latitude of Chaco (36 degrees), it reached an altitude of 15.5 degrees above the southern horizon. The construction of as many as nine great houses began in late 800s, such as those in the "downtown" core of Chaco, Peñasco Blanco, Pueblo Bonito, Hungo Pavi Pueblo, and Una Vida. Especially the residents of Peñasco Blanco, high on the southern rim of the canyon could not have missed seeing the spectacular new star. On May 6, 1006, the super-nova appeared in the southeastern skies of Chaco soon after sunset and reached its greatest altitude around 10:00 P.M. due south.

Astronomical Magnitudes

Sun −26.7
Full moon −12.6
1006 supernova −9.5
1054 supernova −5
Brightest Venus −4.7
Brightest Jupiter −2.8
Brightest nighttime star (Sirius) −1.5

Faintest star visible with unaided
human eyes in dark skies 6
Faintest star detected with the

I sometimes wonder how much this brilliant star, which appeared suddenly out of the darkness and hanging low on the southern horizon, acted as a lodestone for the Chacoans. They came from the north and per-haps, Magi-like, they were drawn southward. Archaeologist Steve Lekson has written about the strange power of the north-south meridian that passes through Chaco Canyon, and some of its power may have come from the exploding star.[4] After the fall of Chaco around 1135 CE, power moved to Aztec 53 miles to the north, then after the fall of that culture, Lekson suggests some migrants traveled to establish the city of Pacquime some 390 miles to the south. The similarities of longitude of these sites are puzzling. Is it a coincidence or did they have the ability to establish a north-south line far beyond their horizons?

Chaco Meridian

Peñasco Blanco 108°00'12.7"
Aztec Tri-Wall 108°00'00"

Nearly fifty years later, on July 4, 1054, another supernova appeared in the skies above Chaco, visible in the daytime for twenty-three days and present in the dark skies for nearly two years. It faded from sight, according to Chinese records, on April 16, 1056. Reaching a magnitude of –5, brighter than Venus, it was significantly fainter than the 1006 CE supernova. This "guest star" was extensively observed in China and Japan. Occurring during the peak of Chacoan culture, it could not have been missed. This explosion produced the famous Crab Nebula with a rapidly spinning pulsar in its center, one of the most intensely studied objects in astrophysics.

Twelve years later, in 1066, Halley's Comet appeared, frightening Europeans on the eve of the Battle of Hastings. The comet was first visible in the east in early April. After briefly disappearing in the dawn, it reappeared on April 24 in the northwest, lasting for some sixty-seven days. It was viewed as an evil portent by the Saxons. How the Chacoans must have puzzled over the mysterious apparition!

In 1077 sunspots large enough to be seen with the naked eye were reported in China, beginning a more than 200-year period of unusual sunspot activity. On July 11, 1097, a total eclipse passed over the Southwest. The inhabitants of Chaco Canyon may have been so startled and puzzled by these events that they became devoted sky watchers, investing more effort in astronomy than they might have had the heavens been ordinary and unchanging.

Construction in the Canyon

Some 300 years before the migrations from north of the San Juan River occurred, there were two large Basketmaker III communities.[5] Their great kivas have tree-ring dates of 580 CE at Shabik'eschee, with seventy pit structures to the east of the Chaco core, near Wijiji. The Basketmaker III site at the western boundary of the park, near Peñasco Blanco, 29SJ423, has more than 100 pit structures. Contemporary sites elsewhere in the Southwest had many fewer structures. Chaco Canyon started early in becoming an interesting place and attracting people who wanted to build large communities.

The first great houses to be built in the Chaco core were Peñasco Blanco, Pueblo Bonito, and Una Vida. They were extensively modified over the next two centuries, but their foot points were put in place between 860 and 900 CE, establishing the basic geometry of great houses in the core of Chaco. Other early great houses built in the late 800s are

Great House and Great Kiva	Approx. Dates of First Construction (CE)
Early Bonito Phase	
Una Vida	860–865
Peñasco Blanco	900–915
Pueblo Bonito	860
Pueblo Pintado	875–925
Classic Bonito Phase	
Chetro Ketl	1010–1030
Pueblo Alto	1020–1040
Casa Rinconada	1027–1054
Pueblo del Arroyo	1075
Late Bonito Phase	
Wijiji	1110–1115
Tsin Kletzin	1110–1115
Kin Kletso	1125–1130
Casa Chiquita	1100–1130
New Alto	1100–1130

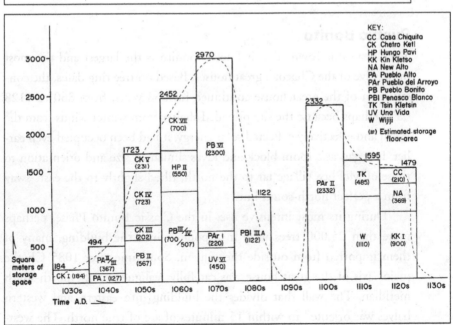

FIGURE 5.2. Construction of storage areas in great houses.

Kin Bineola, some 11 miles to the southwest of the Chaco core, and Pueblo Pintado, some 17 miles to the southeast.[6] It would be extraordinary if the placement of these early great houses revealed a knowledge of astronomy beyond that carried by migrants who came down from north of the San Juan River. For example, awareness of the lunar standstill cycle probably came much later, and it seems unlikely that these early great houses were built along lines established by the major or minor lunar standstills as suggested by Sofaer.[7]

The next two major construction phases are illustrated in Figure 5.2. The first phase started about 1020 CE and marks the onset of the social, political, and religious transformations that led to the establishment of the great Chacoan regional system. It was probably during this period that there was a great growth of astronomical knowledge. However, the great houses that were built in the Late Bonito Phase also have intriguing astronomical features. Wijiji was built at a place with a southeastern horizon that has a remarkable natural marker for the date of winter solstice. Tsin Kletzin lies on a very precisely established north-south line from Pueblo Alto. Kin Kletso provides another marker for winter solstice. New Alto was built along the north-south line of the great kiva of Casa Rinconada.

Pueblo Bonito

The "Beautiful Town" that is Pueblo Bonito is the largest and the most impressive of the Chacoan great houses. Based on tree ring dates, the construction of the great house continued for 268 years, from 860 to 1128 CE. Perhaps because the site provided shelter from winter winds from the north and effectively collected solar energy, it had been occupied even earlier. It began as a room block and kivas similar in size and orientation to those of McPhee village far to the north, tilted slightly to the east, away from a perfect north-south line.

During its most intensive use, in the Classic Bonito Phase, perhaps more than 25,000 trees were incorporated into the building, many of them imported from outside the canyon. Sometime after 1085 CE the major axis of the great house was carefully realigned to the north-south meridian. The wall that divides the building into eastern and western halves was oriented to within 15 minutes of arc of true north. The western half of the south wall was oriented within 8 minutes of arc of true east.

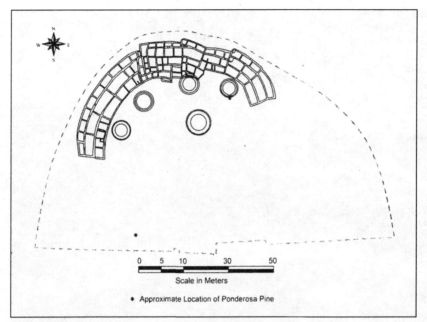

FIGURE 5.3. Pueblo Bonito, construction stage I, 850–935 CE.

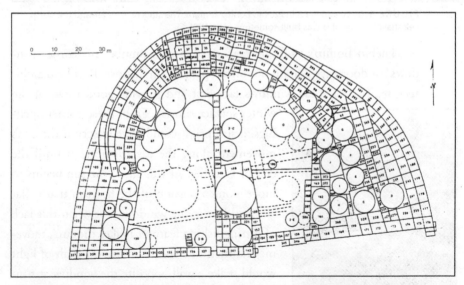

FIGURE 5.4. The Great Kiva A at Pueblo Bonito, showing that the major axis was carefully realigned to the north-south meridian.

The great kiva in the eastern half of the Pueblo departs from north by only three-fourths of a degree east of true. These structures that are carefully aligned to north appear to have been built during the last major building stage of the pueblo.

FIGURE 5.5. Great Kiva A in Pueblo Bonito at night. The stars revolve around the north celestial pole. The kiva was built sometime after 1085 CE.

Pueblo Bonito's walls are perforated with a number of "corner windows" or doorways that do not often appear in prehistoric Puebloan architecture. Two of these windows face east and provide good views of the

FIGURE 5.6. Sunlight entering a corner window in Pueblo Bonito at winter solstice.

winter solstice sunrise. From these corner openings, a sun priest could have observed the sun's movement along the eastern horizon until the end of October. At this time, the sun begins to move along a portion of the horizon that is flat as viewed from the window. Just when this lack of distinguishing features makes the sun's movement difficult to discern, a narrow shaft of light would strike a wall opposite the window at sunrise. As winter solstice approaches, the beam widens and travels northward along the wall. At the solstice, the sun neatly throws a square patch of light into a corner of the room.

Because the movement of the beam along the wall is so obvious from day to day, the corner window at Pueblo Bonito could have been used to predict the solstice as well as to confirm it. A sun priest may have marked the plaster along the wall to keep track of the number of days remaining until the sun arrived at its winter "house." Anticipating important dates may actually have been the most crucial of the sun priest's duties; elaborate ceremonies in honor of the sun would take several days, even weeks, of preparation. Unfortunately, we cannot be absolutely confident that the corner openings in Pueblo Bonito were intentional sun stations because of uncertainties about the building's reconstruction. Outside walls may originally have blocked the views of the horizon and prevented the sunlight from entering the rooms.

Peñasco Blanco

Of the four great houses that had their beginnings before the eleventh century, Peñasco Blanco is uniquely established on the canyon rim, with an unobstructed view of the sky and dramatic views southeastward into the canyon. Since the first construction phase was 900–915 CE, residents of Peñasco Blanco would have easily seen the two supernovas of 1006 and 1054 CE, as well as Halley's Comet.

FIGURE 5.7. "Supernova" pictograph.

The so-called supernova pictograph is found on the canyon walls just north of the great house. On the underside of a low overhang, a Chacoan artist used red paint to depict a sun, a crescent moon, and a star, perhaps signing his work with a hand print.[8] A lively debate has revolved around the depicted conjunction of the crescent moon with the large star. Although the morning star and crescent moon combination is a favored motif in Puebloan mythology and art, it was initially identified, rightly or wrongly, with the 1054 CE supernova that resulted in the formation of the Crab Nebula. The

FIGURE 5.8. Possible pictograph of Halley's comet.

supernova would have first appeared close to a waning crescent moon.

However, anthropologist Frank Hamilton Cushing, who lived with the Zuni Indians for nearly five years in the late 1880s, described an almost identical set of symbols painted on a pillar marking a Zuni sun-watching station with no con-

FIGURE 5.9. Saxons viewing Halley's Comet in alarm in 1066 CE (Bayeux tapestry).

nection to the supernova.[9] Anthropologist Florence Halley Ellis argued that the Puebloans usually did not record unusual events and believes these pictographs were clan symbols.[10]

Actually, the most intriguing pictograph lies on the vertical wall beneath the moon, star, and hand images. Consisting of concentric circles and apparent flames extending to the right, this faded pictograph could well be an image of Halley's Comet of 1066 CE, which first appeared in the constellation of Pegasus on the morning of April 2, according to Chinese, Korean, and Japanese observers. As it moved eastward it was lost in the glare of the sun. On April 24 it reappeared in the northwestern skies, initially without a tail. Its tail soon reappeared and stretched far across the sky, reaching the area of the North Pole. The comet disappeared after sixty-eight days. It was viewed as an evil portent by the Saxons, and the Bayeux tapestry shows a terrified King Harold viewing the comet. It seems to me that the pictograph below Peñasco Blanco is a better depiction of a comet than that on the Bayeux tapestry.

Casa Rinconada

FIGURE 5.10. Waiting for the dawn: June solstice in Casa Rinconada (visitors are no longer allowed inside the great kiva).

On the south side of Chaco Wash, almost directly across from Pueblo Bonito, lies the great subterranean kiva that is Casa Rinconada. With a diameter of over 63 feet, it is one of the largest kivas ever built by the Ancestral Pueblo peoples. Unlike most of those kivas, Casa Rinconada stands alone, physically unconnected to any great house. It is not clear

FIGURE 5.11. June solstice sunlight entering the western niche of Casa Rinconada.

whether Casa Rinconada served as a communal religious structure for the villages near it, a meeting place for residents of the canyon who lived on both sides of Chaco Wash, or a site for large festivals involving both residents and visitors. The unusual underground passage in the northern side of the kiva suggests that reenactments of creation mythologies may have been performed there, as dancers emerged from below the floor.

Whomever it served, Casa Rinconada was constructed with great attention to detail, and, as already noted, it may have been a symbolic representation of the Ancestral Puebloan cosmos, uniting the worlds below with the heavens above. One axis of symmetry was established by the line connecting the north and south doorways; the axis is within one-third degree of true north. The New Alto great house lies along this line, less than 26 minutes of arc from true north. This alignment can be easily verified by sighting from the southern doorway.

Twenty-eight niches ring the kiva's walls; the line from the eighth to the twenty-second niche was also very closely aligned with true east-west, within 8 minutes of arc. A twenty-ninth niche may have been lost in the reconstruction. These niches may have been related to the 29.5-day lunar month.

The remarkable accuracy of alignment of this large and complex structure to the true cardinal directions was demonstrated by Ray Williamson.[11]

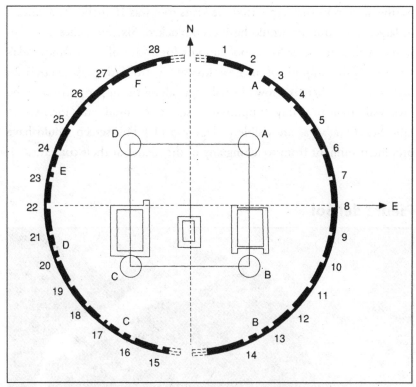

FIGURE 5.12. Casa Rinconada. Niche E is illuminated by sunlight at summer solstice dawn.

The care with which Casa Rinconada and portions of Pueblo Bonito have been built in parallel with the larger cosmos offers a powerful insight into the symbolic world of the Chacoans. The sockets for the four roof supports form a square, each side within one-half degree of east-west and north-south; the western and eastern pairs have orientations of 0 degrees, 29 minutes of arc, and 359 degrees, 24 minutes of arc, respectively. These posts may also have symbolized the directions of the sunrise and sunset at summer and winter solstices.

In addition to the twenty-eight niches, there are six larger and less regularly spaced wall crypts. During the summer solstice sunrise, a beam of light enters a northeastern opening in the kiva and settles into niche E. Although it is a very popular sight among summer visitors to the canyon, it may not have been viewed by Chacoans. Stabilization and reconstruction at Casa Rinconada has not been sufficiently well documented to guarantee that today's effect actually occurred in the eleventh century. The walls of a room outside the northeastern opening that allows sunlight to enter the kiva may have been closed to the outside. Furthermore, if the northeastern post that supported the kiva's roof was 16 inches in diameter or larger, the sunbeam would have been blocked. Sixteen inches is not an unreasonable diameter for a roof support for a kiva of Casa Rinconada's size; one of the supports of the great kiva at Chetro Ketl measured over 26 inches. Finally, there is a row of small vigas along the upper portion of the kiva wall that probably supported a screen of mud and interwoven branches. (The vigas are visible in Figure 5.11.) This screen would have prevented sunlight from touching any of the niches or their contents.

Piedra del Sol

FIGURE 5.13. South face of Piedra del Sol.

Whoever produced the solar spiral on Fajada Butte had to know the days of the solstice from an independent source. Because of the smooth horizon visible from Fajada Butte, the date of the summer solstice needed to be established somewhere else. A spiral petroglyph on the northeastern face of Piedra del Sol may have been just such a source and the date of solstice could have been directly communicated to a person on Fajada Butte.

FIGURE 5.14. (top) Northeast face of Piedra del Sol with theodolite. Note the spiral petroglyph. **FIGURE 5.15.** (right) Spiral petroglyph closeup.

The northeast face contains a spiral petroglyph with a diameter of 17 inches and thirteen to fourteen turns. The center of the spiral is some 5 feet above the present ground level, and its top arm reaches vertically to a height of 8 feet. Approximately two weeks before the time of summer solstice, a pyramidal rock located on the northeastern horizon at a distance of 225 feet casts a triangular shadow across the center of the spiral. Because of the diffuse penumbra of the shadow, the position of the shadow on the spiral petroglyph cannot be used as a precise day marker. However, an observer positioned just in front of the center of the spiral can make precise, direct observations of the sun on the horizon.

FIGURE 5.16. Shadow on spiral petroglyph on June 5. Note the vertical pecked arm of the spiral at the top of the shadow.

MAY 24   JUNE 5   JUNE 20

FIGURE 5.17. Sunrise as viewed from the center of the spiral petroglyph.

As viewed from such a location, the sun rises over the top of the pyramid on June 4–6, providing fifteen to seventeen days of prediction. Because the top of the pyramidal rock has an altitude of 14.36 degrees as viewed from the center of the spiral, the effects of varying atmospheric refraction are small, and precise calendrical observations would have been possible. Before June 5, successive dawn suns climb the southern edge of the pyramid on the horizon, giving further opportunities for prediction of the solstice. For a few days around solstice, the sun rises on a clearly identifiable notch to the north of the pyramid, providing the opportunity for confirmation of that date. There is a direct line of sight between Piedra del Sol and the three slabs of Fajada Butte, and observations made at the spiral of Piedra del Sol

FIGURE 5.18. South face of Piedra del Sol.

could have provided the information necessary for the larger spiral to be located properly on the cliff face behind the three slabs.

Two humpback flute players, a bugle player, two animal figures, and an unusual circle with curved rays have been pecked upon a 6.5

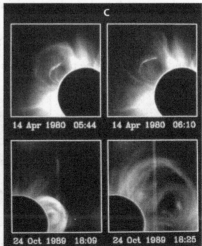

FIGURE 5.19. A) Petroglyph on southern face of Piedra del Sol that may depict the eclipse of July 11, 1097, the upper left may be a representation of Venus; B) drawing of the eclipse of July 18, 1860, which apparently shows a coronal mass ejection in progress; C) photographs of two coronal mass ejections.

by 10 foot area on the south face. The image looks like the curved coronal structures of the eclipsed sun during a coronal mass ejection or exaggerations of curved polar streamers of the corona. The eclipse of July 11, 1097, was the only total solar eclipse to cross the canyon during the period of maximum great house construction (Figure 5.2). Occurring in mid-afternoon with a duration slightly greater than four minutes, when the sun was high in the sky with an altitude of 58 degrees, the eclipse was well placed for observation. Since it occurred close to the beginning of a period of very unusual sunspot activity associated with the so-called Medieval Maximum, the sun at that time may have been in a state of enhanced activity with a coronal mass ejection in progress.

A pecked circle to the upper left may represent the planet Venus, which was the brightest object in the sky during the eclipse, lying 34 degrees east of the sun with a magnitude of −4.0. The sudden appearance of brilliant Venus during a total eclipse often is a spectacular and dramatic phenomenon. Admittedly, the proposed connection between the eclipsed sun and Venus is highly speculative, but no more so than those involving the Taurus supernova or Halley's Comet on the panel below Peñasco Blanco. The Piedra del Sol site has the advantage that it may have been associated with a systematic program of sun watching at the spiral on the northeastern face. When the sun returned to the top of the pyramid around July 6–8, less than a week before the date of the 1097 eclipse, sun watchers may have been in the vicinity.

A pecked basin with a diameter of 3 inches and a depth of 2 inches is contained on the detached shelf in front of the west face of the rock. The rounded shapes of the pecked basins similar to those found in Mesa Verde

FIGURE 5.20. South face of Piedra del Sol: left: pecked basin on detached slab; right: grinding area behind slab.

FIGURE 5.21. Sunset December 7.

(Figure 8.2) distinguish them from the more common straight-sided bedrock basins associated with Chacoan stone circles (Figure 5.24). The basin can be approached by means of steps cut diagonally across the face of the rock. As viewed from the basin, the sun sets on the southwestern horizon behind an isolated butte approximately 13–14 days before the winter solstice or close to December 7. Behind the shelf a narrow corridor provides access to a bedrock grinding feature, also similar to those found associated with pecked basins at Mesa Verde. This bedrock metate has a length of 12 inches and a width of 4.75 inches and is set on a slightly downward-sloping ledge that has itself been ground smooth.

The three faces of Piedra del Sol have different astronomies, which were probably established at different times. The spiral on the northeast face may have been a primary calendrical station for summer solstice and may have been used to establish dates for periodic festivals and trade fairs during the Bonito Phase. The petroglyph on the south face may represent the total eclipse of 1097, which occurred at the start of the Late Bonito Phase. The western face appears to be associated with the late reoccupation of the canyon in the late twelfth and thirteenth centuries. Besides marking a place from which to view the setting winter solstice sun, the pecked basin on the western shelf may have been used as a repository for offerings to the sun, consisting of water or ground material (corn, shell, or semiprecious stone). Similar grinding features and pecked basins are found at the Gallo Wash site near the campground. The Gallo site is associated with McElmo masonry, and its grinding areas were thus probably produced during the reoccupation of the canyon in the McElmo Phase (1140–1200 CE) and/or the Mesa Verde Phase (1200–1300 CE).

Wijiji

At the eastern extreme of the settlement in Chaco Canyon stand the ruins of what may have been the last of the great houses to be built in the canyon. Here, at the very edge of the "world's center," the Ancestral Puebloans may have watched the sun rise and set at winter solstice, the harbinger of snow and harsh winds, with fervent prayers for its northward return and a warming of the earth. A little beyond the great house of Wijiji, a staircase leads to a ledge along the canyon rim.[12] A faded sun symbol, probably painted by Navajo Indians, marks the site. Farther along the ledge are some boulders, one of which is marked with crosses and spirals, perhaps cut by the Ancestral Puebloans. From just a few yards beyond this spot, a sun priest could have observed the winter solstice sun rising from directly behind a natural rock pillar on the other side of a bend in the canyon rim. And at sunset on the same day, the sun sets into a natural cleft in the rock ledge a short distance away.

The great house itself, built in a single construction effort around 1110 CE, provides an excellent primary calendrical station for winter solstice. But even before construction of the great house, the area may have been a calendrical station for winter solstice. As viewed from the northwestern corner of the house, the rising sun drops into the sharply defined northern edge of the notch on the southeastern horizon on December 4–5, thereby providing a sixteen- to seventeen-day opportunity for precise prediction of winter solstice. During the next two-and-a-half weeks, the rising sun moves across the notch and reaches the southern edge at winter solstice. This remarkable behavior of the sun just around solstice may have been partly responsible for the particular location of that great house.

Pueblo Alto

The trail to Pueblo Alto passes Kin Kletso, which may have functioned as another calendrical station. The great house was constructed between 1125 and 1130 CE, rather late in the Chaco regional system. It provides

NOVEMBER 27

DECEMBER 4

DECEMBER 5

DECEMBER 22

FIGURE 5.22. Sunrises as viewed from Wijiji between November 27 and December 22. The "fall" of the sun into the left hand side of the gap provides a 16-day anticipation of winter solstice, which is confirmed when the sun reaches the right hand side of the gap. This anticipation period is similar to that at Piedra del Sol for summer solstice and Kin Kletso at winter solstice.

FIGURE 5.23. Kin Kletso sunrise December 22.

another sixteen- to seventeen-day opportunity for prediction, when standing on the south wall, and confirmation, when standing on the north wall. The winter solstice sun "falls" into the base of the southwestern cliff at those two dates. In a manner similar to that at Wijiji, throughout the fall the sun rises farther and farther south along the horizon above the cliff. On the December solstice it rises just at the base of the cliff, as seen from the

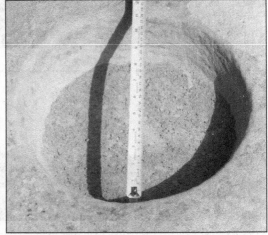

FIGURE 5.24. Round and square bedrock basins on the northern rim of Chaco Canyon. The two basins in 29SJ1565 (left) are the largest in the canyon and lie on line connecting Pueblo Alto and Tsin Kletzin. Rectangular basin: 18x24.4x4.7 inches; circular basin: 18.1x20x5.9 inches; (right) closeup of a round basin.

northern side of the great house. The large boulder west of the ruin may have served as a sun watching station before construction of the great house.

After reaching the rim of the canyon through the narrow cleft behind Kin Kletso, one walks eastward along the rim toward an excellent viewpoint of Pueblo Bonito. Along the way there are several circular basins carved in bedrock. These beautifully created basins are often enclosed by stone circles, and were first studied by Tom Windes in the 1970s.[13] Were they meant to hold water offerings or sacred objects? Sight lines to other features of the canyon or to the sun may be important. According to Windes, each of the basins has a view of a great kiva. These basins remain some of the great mysteries of Chaco.

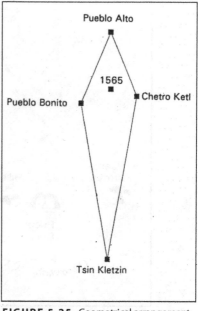

An unusual stone circle, 29SJ1565, (Figure 5.24) lies farther along the canyon rim, beyond the trail to Pueblo Alto. It has been placed on the north-south line connecting Pueblo Alto and Tsin Kletzin on the south rim of the canyon.

One of the few such basins outside Chaco Canyon is found at Chimney Rock, further evidence of Chacoan influ-

FIGURE 5.25. Geometrical arrangement of Tsin Kletzin, Pueblo Alto, Pueblo Bonito, and Chetro Ketl. The large stone circle with bedrock basins is on line connecting Pueblo Alto and Tsin Kletzin.

ence at that outlier (Figure 6.16). The basin has a view of a great kiva, but perhaps more meaningful is that the basin marks the spot from which one can observe June solstice sunrise along the north wall of the Chimney Rock Great House. Is it possible that other basins in Chaco also have astronomical meaning?

The first construction phase at Pueblo Alto, which occurred at approximately 1020 CE, established the orientation of the north wall, which is some 1.1 degrees off an east-west orientation. It is the earliest and least accurate of the north-south alignments of the great houses and may have been constructed before Chacoans had refined the technique of shadow casting of a gnomon. To the west is New Alto, which lies within 26 minutes of arc of the north-south axis of Casa Rinconada. Built sometime after 1100 CE, New Alto was sited after the great kiva had been

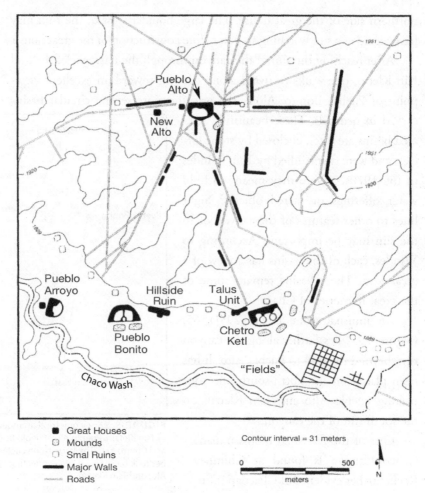

FIGURE 5.26. System of roads and walls around Pueblo Alto.

built with its accurate alignment. Someone standing at the southern entrance of Casa Rinconada could have established the location of New Alto by a signaling system.

Fajada Butte

The sun dagger of Fajada Butte has become one of the great icons of Chacoan astronomy. Publicized by the film, *The Sun Dagger,* narrated by Robert Redford, a "dagger" of light formed by two slabs leaning against a cliff once passed through the center of an elegantly pecked spiral on the day of the summer solstice. In June 1977 rock art specialist Jay Croty and artist Anna Sofaer were recording spiral petroglyphs near the butte's summit for a rock art survey. Near midday, they observed a dagger of light moving across the rock wall behind three slabs. Since the dagger appears for only fourteen minutes, they were very lucky to be at the right place at the right time.[14]

The sandstone slabs are approximately 6 to 10 feet high, 2 to 3 feet wide, and 8 to 20 inches thick. Each of the slabs weighs approximately 4.5 tons. These slabs were apparently part of a single block that broke off the cliff, toppled over, and split along bedding planes. They are now separated by some 4 inches. Since the time of their discovery, they have shifted slightly so that the dagger of light no longer passes through the center of the spiral on the June solstice.

A mysterious feature of the butte is the massive ramp that was built on its southwestern side. The ramp has a length of approximately 700 feet and rises 280 vertical feet.[15] Its base may connect with a road segment that is associated with a great kiva in a nearby rincon. The ramp stretches from the toe of the talus at the base of the butte to its top. Its function could be similar to that of stairways in the canyon, which may be more symbolic than practical, namely expressive of shamanic themes of ascent and descent. Stairways are symbols of travel between the worlds below and above. This ramp seems more private than other stairways in the canyon because it is uniquely associated with the local community near Fajada Butte.

Rooms at the Top

The ramp also provided access to approximately fifteen rooms built along 100 yards of the western portions of the first terrace beneath the top of the butte. A catwalk may have connected these rooms to another ledge on the eastern side, where there are another twenty or so rooms along its length of more than 130 yards. In addition to these thirty-five rooms, there is one circular kiva, suggesting

FIGURE 5.27. Chaco archaeologist Dabney Ford pointing out the rooms and walls on the eastern ledge of Fajada Butte.

FIGURE 5.28. Walls on the eastern ledge of Fajada Butte.

that the site was more than simply a desperate overnight hiding refuge. The presence of manos, remains of cooking pots, burned corncob, and hearths indicate these remote and inaccessible rooms were used as residences. The rooms are built against the cliff, such that the cliff forms the back wall of each of the rooms. The ceramics associated with these rooms primarily date from the 1200s, or the post-Chaco reoccupation of the canyon.[16]

These two sites are a great puzzle. When did people live there, and why would they have chosen such a remote and precarious place for building residences? Regular human presence on the summit would have facilitated discovery of the light and shadow effects on the wall behind the three slabs and eventually the recognition that they could be used to mark the summer solstice. The wall construction of the room does not contain decorative elements in the form of chinking stones, such as those found in the great houses down in the canyon; nor is there any of the core-veneer construction that is so distinctive of the Bonito period when the great houses were constructed. It seems unlikely that the rooms were intentionally built during the Bonito Phase to provide habitation for astronomer priests to visit the spiral.

The rock art in the vicinity of the rooms was reached by the roofs and thus provides some clue to its dating. There are spirals, intertwined spirals, a snake, and two rectangular figures over which shadows pass throughout the year.[17] The various petroglyphs are similar to those found at Hovenweep National Monument in the vicinity of Holly House, which probably date to the 1200s.

The Puzzle of the Sun Dagger

There must be a fascinating story behind the three-slab site and the sun dagger. Who first found the light and shadow effects on the wall behind the slabs? Who discovered that they could actually be used to mark June solstice? When were the petroglyphs put in place?

On the June solstice, the dagger of light produced by two of the three slabs passes across the rock wall behind the slabs for only fourteen minutes before noon, starting at 11:07 A.M. local solar time. Because it is so fleeting and doesn't occur at sunrise, it is likely that such an event could only have been discovered by people living near the top of the butte. Someone wishing to greet the sun at dawn would have missed the effect entirely. We know from the Mesa Verde black-on-white ceramics found in the rooms at the top that people were living near the top of the butte in the 1200s, probably seeking safety during unsettled and dangerous times. The rock artist who produced the spiral on the rock wall behind the slabs could have obtained the date of the June solstice from Piedra del Sol, which is directly visible from the three-slab site.

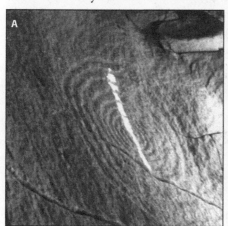 

FIGURE 5.29. The three-slab site on Fajada Butte: A) the sun dagger near June solstice. Note the diagonal pecked line that passes through the center of the spiral. B) The three-slab site.

The purported markings of major and minor lunar standstills are much less certain than the light and shadow effect at June solstice.[18] If the solar spiral was produced during the reoccupation of the canyon in the 1200s, major lunar standstills had already been discovered at Chimney Rock (the East Kiva at the Chimney Rock great house was built in 1076 CE). Major lunar standstill sunset appear to have been marked in Mesa Verde at Cliff Palace, between 1186 and 1261 CE (Figure 8.9). Some of the post-Chaco residents of the canyon probably came from the Mesa Verde area and may have learned about lunar standstills at Cliff Palace and the Sun Temple. They would have known about the major standstill of the moon. However, the minor standstill (9.3 years later) is a much more subtle phenomenon. Sofaer[19] and her colleagues suggest that it may be recorded on the spiral by a diagonal line crossing the center at the same angle as a shadow produced by the moon near the horizon. That pecked line is not as carefully produced as the spiral itself and appears to have been produced at another time, which actually would not be surprising as the northern lunar standstill cannot be seen near the date of the summer solstice. The most likely time for someone to discover a shadow of the moon crossing the center of the spiral would have been the full moon rising at sunset close to the next winter solstice.

However, the line through the center of the spiral may not mark the moon at all, but rather sunrise a few weeks before or after the June solstice, when the sun has a declination of 18.4 degrees around the middle of May and end of July.[20] The shadow cast by the easternmost slab at sunrise is easily observed and easily marked. The diagonal line through the center of the spiral may thus have nothing to do with the moon. The absence of a similar line corresponding to the much more observable major lunar standstill argues strongly against any intentional lunar markings on the spiral.

Summary of Chacoan Astronomy

There are many elements in the story of astronomy in Chaco. Some are well substantiated, and some remain highly controversial. The two sources of Chacoan astronomy appear to be migrations of people from the north bringing a belief in the importance of the north-south meridian and the growth of periodic festivals and trade fairs during the Chaco fluorescence. Interest in the north-south meridian may have been intensified by the appearance of supernova 1006 in May of that year. The need for precise timing of trade fairs and periodic festivals could best be satisfied by a horizon calendar, which required careful sky watching at calendrical stations such as Piedra del Sol and Wijiji. The Chacoans developed an ability to establish the north-south meridian with extraordinary precision, which they demonstrated at Pueblo Bonito, Tsin Kletzin/Pueblo Alto, and Casa Rinconada. June solstice is beautifully marked at Fajada Butte, but it seems likely that the solstice petroglyph on the three-slab site was not established by Chacoans but rather during the reoccupation of the canyon in the thirteenth century.

The cycle of the sun from one solstice to another is easy to establish. The 18.6-year standstill cycle of the moon is less easy to observe. The standstill cycle may have been discovered at Chimney Rock and first acknowledged by Chacoans in 1076 CE with the construction of a kiva on the high mesa below the chimneys. However, there is no strong evidence in the archaeological record of Chaco Canyon itself for knowledge of lunar standstills. Lunar standstills are probably not marked at the three-slab site of Fajada Butte, nor does it seem likely that any of the great houses were intentionally aligned along lines of major or minor standstill.

CHAPTER 6

CHIMNEY ROCK

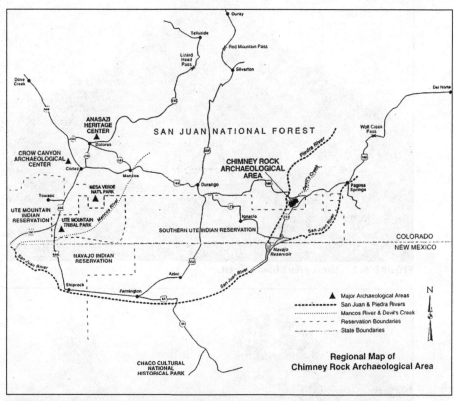

FIGURE 6.1. The neighborhood of Chimney Rock.

There can be few travelers along the highway between Pagosa Springs and Durango in southwestern Colorado who have not been intrigued by the double spires of Chimney Rock. Looking up from the highway, one can spot the visitors' tower on the high mesa and the ruins of the Chimney Rock Pueblo, the Chacoan great house, which dominated the mesa.

Originally a two-story structure, it must have been imposing, built on the highest accessible spot in the area. What was the purpose of building such a large building so far above water and agricultural land? Was it a fortress, a castle, a cathedral, a monastery, an observatory, or perhaps a trophy home? It was probably a little bit of each.

FIGURE 6.2. Chimney Rock from the east.

FIGURE 6.3. Chimney Rock (right) and the Companion Chimney (left).

FIGURE 6.4. Companion Chimney and south wall of the Chimney Rock Pueblo.

The Astronomy of Chimney Rock
Places to Visit

Solstice sunrise: Sun Tower
                              Bedrock basin
                              Sunrise shadow viewed from the Great House

Equinox sunrise: Peterson ridge (permission required)

Major standstill moonrise:
                              Observation Tower
                              Guard House

Nine centuries ago the view from below would have been even more remarkable. The two-story Chimney Rock Pueblo, gleaming in the sunlight, perhaps plastered and painted white, would have been a stunning sight to those walking along the river canyons. Chimney Rock was home to the most northeastern segment of the Ancestral Pueblo peoples. Lying some 93 straight-line miles to the north of Chaco Canyon, the Chimney Rock Pueblo is the loneliest and most isolated of all outliers (45 miles to Aztec Ruin), the highest (7,600 feet), and the most removed from arable land (1.24 miles).[1] The area contains eight communities, some of which date from Pueblo I times.[2] The chimneys and the associated high mesa have been recognized by the Taos Pueblo of northern New Mexico as a shrine dedicated to the War Twins of Pueblo mythology.[3]

Before 1050 CE, there had been a slow but steady leapfrog movement of people northward from the San Juan River along the Piedra River. Clearing vegetation along the riverbanks would have caused a gradual entrenchment of the river, a lowering of the water table, and a reduction in agricultural yields. Hunting would have reduced the numbers of deer and elk in the area. The combined effects of poorer agriculture and hunting would have driven people

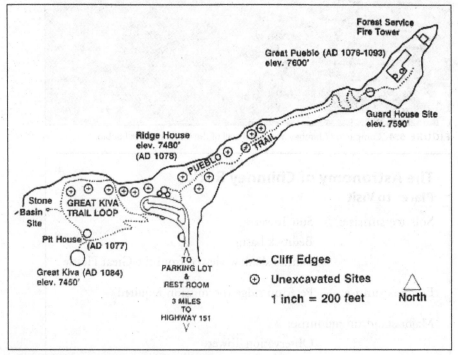

FIGURE 6.5. Trails and archaeological features at Chimney Rock.

upstream, and they would have become increasingly isolated as they progressed farther from the San Juan River. Their northward movement was stopped by a growing season that was too short for maize agriculture along the Piedra River near Chimney Rock. Of those who reached that area of the Piedra River, some settled on the high mesa of Chimney Rock in the early part of the decade of 1050 CE, perhaps to escape from the cold air pooling along the river.

Drainage of cold air into the floor of the valley would have resulted in a shorter growing season in the valley floor than on the upland terrain. Corn requires more than 100 frost-free days to mature, and in order to achieve such a growing season, the Ancestral Puebloan farmers in many areas were sometimes forced to move away from the most fertile agricultural lands closest to water in order to get above the pool of cold air. The movement of people into the Chimney Rock area appeared to be gradual through the Arboles Phase (Figure 6.6), and then there was a rapid movement upward onto the higher sites during the Chimney Rock Phase. Those who were responsible for constructing the Chimney Rock Pueblo may not have been driven solely by a search for a longer growing season, but may have been attracted by the unique geographical and astronomical qualities of the high mesa. In particular, the high and isolated place on which the Chimney Rock Pueblo was built is reached only by a narrow causeway of rock, falling off steeply on both sides. Passage to the mesa is blocked by an unusual structure known as the Guard House, which we now interpret as a tower associated with the moon. The space

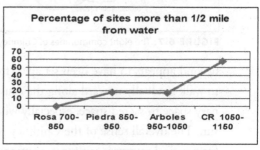

FIGURE 6.6. Location of habitation sites in the Chimney Rock area showing percentage of sites more than 1/2-mile from water.

Archaeological Phases in the Chimney Rock Area

	Dates (CE)	Predominant Arch. Features
Rosa (Early Pueblo I)	700–850	Pit Houses
Piedra (Late Pueblo I)	850–950	Jacal (mud and wattle) Pueblos
Arboles (Early Pueblo II)	950–1050	Masonry Pueblos

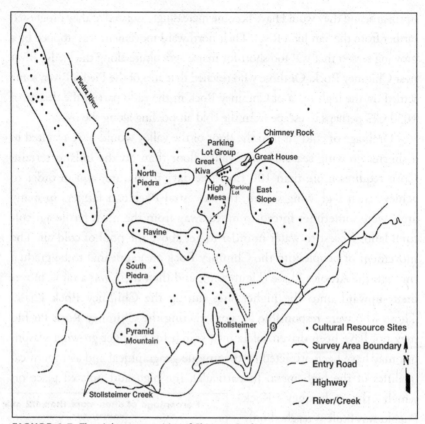

FIGURE 6.7. The eight communities of Chimney Rock.

beyond appears to have been set apart from ordinary activities. Even today, to walk gradually upward along the causeway seems a transition from ordinary space to sacred space, especially when approaching the rising moon or sun. The overall sense of the Chimney Rock Pueblo is that it was not built for practical purposes but for its commanding view of the double spires and the surrounding heavens.

In its final form the Chimney Rock Pueblo contained two kivas and thirty-five ground-floor rooms; a second story may have added an additional twenty rooms to the structure. There was apparently two stages of construction. Based upon tree-ring dates, the East Kiva appears to have been built first in 1076 and the remainder of the great house was probably completed in 1093. The core-and-veneer masonry of the town resembles that of the great houses of Chaco Canyon.[4] The Chaco-style East Kiva contains a horizontal, subfloor ventilator tunnel and vertical shaft, and a 5-foot-high banquette upon which were built eight beam rests.

The East and West Kivas were set within a quadrangle of walls, which had been filled by the prehistoric inhabitants to the level of the kiva roofs, forming courts within the building. Another similarity to Chacoan buildings is the L-shaped layout of the structure, the two arms of which enclose the East Court.[5] A bench has been built along the east-facing exterior wall in the East Court, providing a place to view the double chimneys.

The excavations of the Chimney Rock Pueblo by Professor Frank Eddy of the University of Colorado[6] in the early 1970s revealed that the interior and exterior walls of the pueblo had been placed directly upon bare sandstone bedrock, indicating that prior to construction the upper mesa was a bare rock platform without soil or vegetation. Every piece of the tons of rock and adobe necessary for construction apparently had to be carried up to build the large structure. Since there is no water on the

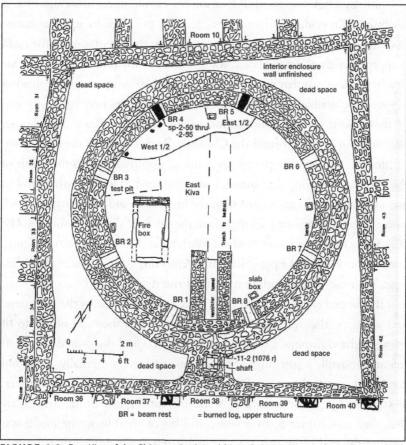

FIGURE 6.8. East Kiva of the Chimney Rock Pueblo, perhaps constructed initially as an isolated kiva in 1076 CE.

mesa, construction may have been carried out during the winter, when snow could have been melted for mixing the adobe mud. But it is doubtful that snow could have provided all the water necessary. We can imagine a steady stream of people working their way upward 1,000 feet from the valley floor, carrying pitch-coated wicker baskets filled with sloshing water.

Lunar Standstills

During the summer of 1988 I organized an archaeoastronomy field school at Yellow Jacket for undergraduate students from the University of Colorado. We investigated the astronomy at a number of sites in the area, including Chimney Rock. Professor Frank Eddy, who led the CU excavation of Chimney Rock in 1970–1972, suggested to me that the high mesa at Chimney Rock was an ideal location for astronomical observations and might contain evidence of ancient astronomy practiced by its inhabitants. Looking over a topographic map, we noted that the orientation of the mesa is approximately along the line toward sunrise on the summer solstice. Our first hypothesis was that we could have spectacular summer solstice sunrises between the double chimneys and that such spectacles may have been one of the reasons for the construction of the Chimney Rock great house. With that in mind, I visited the Chimney Rock Pueblo in the early dawn of June 21. To my disappointment, sunrise occurred well to the south of the chimneys. During that summer we continued to wonder what kind of astronomy might be associated with the high mesa and those majestic rock towers. Venus sometime rises slightly to the north of the sun at solstice. The War Twins may have represented the morning and evening aspects of Venus, and we checked out the possibility that Venus might rise between the chimneys. That didn't work either; it couldn't rise that far north.

By the end of July our team had surveyed the outline of the chimneys, and we knew the necessary coordinates for an astronomical object to fit between the chimneys. I checked out the moon and discovered that at its major standstill, it just might fit between the towers. Actually, I was not too happy about the possibility of getting caught up with the controversies involving major moonrise standstills at Fajada Butte, which are described in Chapter 5. By a wonderful bit of good luck, the moon was nearing the end of its two years of major standstills during that summer of 1988. I calculated that—possibly, just possibly—in the early morning

FIGURE 6.9. Moonrise August 8, 1988.

FIGURE 6.10. Moonrise December 26, 2004.

hours of August 8, the moon might just fit between the towers. My students were skeptical that it would be worth staying up until two in the morning on that high mesa based on my calculations, but they joined me, and we set up a bevy of cameras to catch the effect. Fortunately, the moon behaved beautifully. Everyone was stunned. I felt like a Yankee in King Arthur's court. The moon came up as predicted, we captured its spectacular rise on film, and the rest, as they say, is history.[6] The next time the rising moon was photographed between the chimneys was in the fall of 2004, some sixteen years later (Figure 6.10).

The dates of full moon rise between the chimneys in the latter half of the eleventh century were 1056–1057, 1075–1076, and 1093–1095. These moonrises were at sunset, close to December solstice, and would have been the most spectacular events. When I compared these dates with the tree-ring dates obtained by Eddy from logs found in the Chimney Rock Pueblo, I was astonished to discover that the two episodes of construction of the pueblo correspond with the last two lunar standstills of the century. Could the pueblo have been built primarily to observe the moon coming up between the chimneys? Could the spectacle of the moonrise have so captured the attention of not only priests from Chaco, but also the local inhabitants, that it justified the immense effort required for construction on that remote mesa?

Full moon rising between chimneys	Dec 1056
Full moon rising between chimneys	Dec 1057
Full moon rising between chimneys	Dec 1074
Full moon rising between chimneys	Dec 1075
Partial solar eclipse	7 March 1076
Cutting of trees for East Kiva	July-August 1076
Full moon rising between chimneys	Dec 1076
Cutting of trees for great house	July-August 1093
Full moon rising between chimneys	Dec 1093
Full moon rising between chimneys	Dec 1094
Full moon rising between chimneys	Dec 1095
Total solar eclipse	11 July 1097

Perhaps as early as 1056, residents of the Chimney Rock area discovered the full moon rising between the chimneys at winter solstice. That time was just a few years after the appearance of the Crab Nebula supernova, and unusual events in the sky should still have attracted great attention. They may have built a shrine on the upper mesa to celebrate the event. Eighteen years later the full moon again rose between the chimneys at winter solstice of 1074 CE. A ponderosa pine pole taken from the lower and original horizontal ventilator of the East Kiva has been dated at 1076 CE. Analysis of the outer layers of the log suggests it was cut during the growing season of July-August 1076. The moon again rose between the chimneys on December 1076.

We can see "as through a glass, darkly" the likely sequence of events on the Chimney Rock mesa in 1076. The full moon had last been seen rising between the chimneys on December 14, 1057, and it had returned. Its approach throughout the previous few years could have been noted by its increasingly northward movement at moonrise on the horizon, so the event could have been anticipated. There was time for someone, a resident of Chimney Rock or an itinerant trader, to inform leaders at Chaco of the return of the moon to the chimneys. Then the story gets really interesting. Chacoans came up to Chimney Rock during that summer of 1076, trees were cut, and the free-standing East Kiva may have been completed in time for the winter solstice moonrise of December 13, 1076. The summer cutting of the logs, indicated by incomplete growth rings, is an unusual practice for an agricultural group, which usually engaged in winter, off-agricultural-season cutting.

This was not a standard "cookie cutter" kiva since it had to be built on top of the bedrock of the mesa. Were the Chacoans invited by the residents of Chimney Rock or did they impose themselves on the locals? Who cut the trees, carried and shaped the stones, and built the kiva? Were the locals forced to do the work, or were they also so caught up in the majesty of the moonrise that they enthusiastically cooperated with the Chacoans? The new kiva may have been built on top of a previously existing shrine.

The full moon returned to the chimneys in December 1093, and its return had been anticipated. The second set of dates from the East Kiva corresponds to the time when the floor was filled and raised to establish a higher floor with a second horizontal ventilator shaft. The East Kiva and

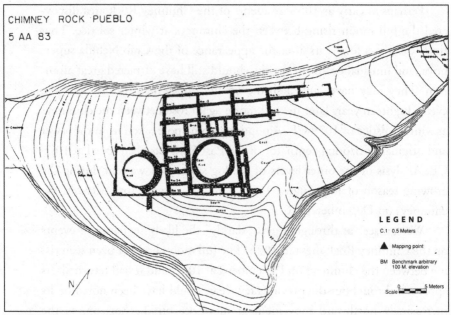

FIGURE 6.11. Chimney Rock Pueblo.

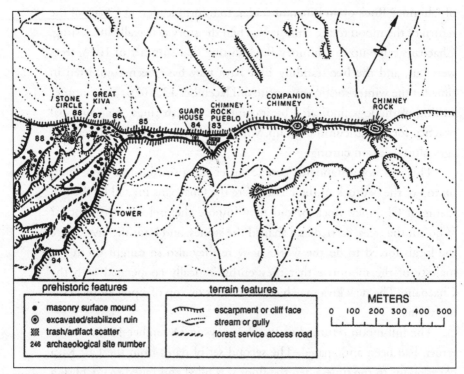

FIGURE 6.12. Map of Chimney Rock, giving the archaeological site designations (e.g., Chimney Rock Pueblo, 5AA83).

adjacent room 8 were roofed over with logs cut in the summer of 1093. The
final combination of two kivas and room blocks is shown in Figure 6.11.

Priests and architects from Chaco must have played pivotal roles in
the design and construction of the great house. It was not unusual for a
Chacoan structure to be built in or near a preexisting community, and we
can only speculate on the dynamics of the interaction between Chacoans
and the residents of the Chimney Rock area. A Chacoan workforce may
have lived in the houses built along the causeway between the main settlement and the upper mesa (sites visible in Figure 6.5 and marked 5AA85
in Figure 6.12), otherwise it is hard to understand why people would have
chosen to build their houses along such a narrow strip of land.

This remarkable agreement in the dates of construction and the
northern lunar standstills does suggest that the great house was built primarily for observations of the moon and associated ceremonies. In the
Chimney Rock Pueblo, in its courts above the kivas, and on its rooftop,
people could have gathered to observe moonrise through the double
spires. It was not necessary for those responsible for planning and construction to have knowledge of the 18.61-year lunar cycle. Anticipatory
observations of moonrise at the start of each northern standstill were clearly
possible by simply observing the position of the moon on the horizon.
The preponderance of remains of deer and elk at the great house indicates
it was primarily used for ceremonial feasting.[7]

The most important astronomical event would have been the spectacle of the full moon rising between the chimneys near the time of winter
solstice. Among the historic Puebloans, sun and moon watching established the ceremonial calendars.[8] The various lunations throughout the
year were given names and were associated with ceremonial activities. The
Zunis attempted to organize their calendar so that winter solstice occurred
at or near the full moon. White Shell Woman (the moon) helped to per-
suade the sun to return north at winter solstice.[9] The coincidence of full
moon and winter solstice would also have provided an important opportunity to bring the solar and lunar calendars into agreement. At Hopi the
sun chief watched the setting sun to establish the date of *Soyal,* the winter
solstice ceremony. In the *Powamu* ceremony, which followed *Soyal,* the
Powamu chief planted beans in a kiva at the time of the full moon.

In both Hopi and Zuni societies, the rising full moon seems to have
been carefully watched near winter solstice. The inhabitants of Chimney

Rock Pueblo may have been similarly watchful of the rising moon. When it rose between the double spires above the snow-covered landscape, colored red from the glow of sunset, the moon must have appeared huge and brilliant. The sight of the moon rising between the chimneys ranks as one of the great dramatic events of the heavens. Our Ancestral Pueblo predecessors, probably standing in the courtyard and on the roof of the Chimney Rock Pueblo, could not help but have been impressed.

The so-called Guard House (5AA84, Figure 6.12; Figure 6.13) was a second place to observe the moon rising between the chimneys. That

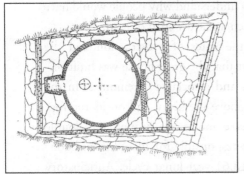

structure has always been something of an enigma, sitting astride the lower neck of the high mesa. Now, it appears it may have been a tower (perhaps similar in intent to the sun tower, 5AA93) designed to obtain a view of the moon rising in the deep gap between the chimneys (Figure 6.14).

FIGURE 6.13. Chimney Rock Guard House.

FIGURE 6.14. Moonrise between the chimneys as viewed from the Guard House.

FIGURE 6.15. June solstice sunrise as viewed near the sun tower.

Summer Solstice

How would Ancestral Pueblo observers have determined the dates of the summer solstice? As seen from the Chimney Rock Pueblo, the sunrise on the distant horizon is blocked by the high chimney for at least a month prior to the solstice. Similarly, the other sites along the top of the mesa have their views of the horizon blocked by the chimneys and the higher portions of the mesa. Where could a horizon calendar have been observed? There is a tower on the high mesa, located above the southeastern cliffs (5AA93). Observations of the rising sun throughout the year could have been easily made from the top of the tower or the canyon rim to the east. At the June solstice the sun rises at a notable dip on the horizon, where

FIGURE 6.16. Bedrock basin near the great kiva.

the near horizon meets the distant horizon. The tower may have been constructed to mark the area for observing the sun. To the southeast of the tower, in the direction of the winter solstice, there is an isolated butte with two kiva depressions on its summit. This may have been a distant shrine for winter solstice.

At the western end of the High Mesa, 164 feet from the great kiva, a circle made of irregular stones surrounds a bedrock basin, 14 inches in diameter, extraordinarily similar to the stone circle and basins of Chaco Canyon (Figure 6.16). The stone circle and basin at Chimney Rock reaffirm its connection to Chaco. The basin marks a place where the two chimneys merge and may have been a place to view the chimneys backlit by the northern standstill moon. As was the case at Chaco Canyon, this basin has a view of a great kiva. The basin also marks the place for observing June solstice sunrise along the northern wall of the Chimney Rock great house. The basin is a secondary calendrical site since the date of solstice needed to be established somewhere else. That date could have been established by the rising position of the sun as viewed from the sun tower. Once the date of solstice had been established, the Chimney Rock great house could have been designed to lie along the sightline from the bedrock basin. A fascinating puzzle of the sightlines from the basin is that the line going to the southern wall of the Chimney Rock Pueblo and the East Kiva aligns with the Taurus supernova of 1054. There would have been a gap of twenty-two years between the appearance of the supernova in the skies over Chimney Rock and the construction of the East Kiva.

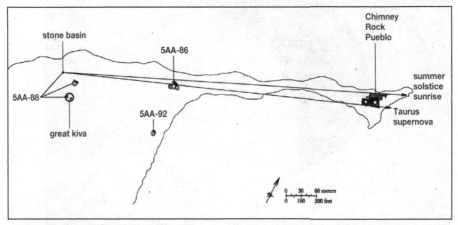

FIGURE 6.17. Sightlines from bedrock basin.

FIGURE 6.18 (top) June solstice sunrise occurring along the north wall of the Chimney Rock Pueblo as viewed from the bedrock basin.

FIGURE 6.19. (middle) Shadow of chimneys at June solstice sunrise—view to the west from the Chimney Pueblo.

FIGURE 6.20. (bottom) June solstice sunrise viewed from Peterson Ridge— view to the east toward the Chimney Rock Pueblo. Note the observation tower to the right of the chimneys, which are merged in this view.

Equinox Sunrise

As seen from the high mesa, the sun never rises far enough north to fit between the pinnacles, but to the west, across the valley on a ridge above the Piedra River, the sun at equinox rises between them twice a year. On that ridge, known as the Piedra Overlook or Peterson Ridge, there are twelve

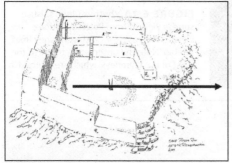

FIGURE 6.21. Overlook Pueblo (5AA8), arrow points to the chimneys.

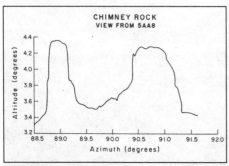

FIGURE 6.22. Profile of the chimneys measured from the Overlook Pueblo.

sites, each of which provides a view of sunrise between the chimneys at a different day of the year.[10] These could have served as "live-in" calendars, in which residents would be assigned to signal to the Chimney Rock Pueblo when they could see the sun rising in the gap between the chimneys.

The largest site of the ridge group, 5AA8, contains three kivas, a two-story room block, and a spectacular view of the chimneys. I visited the ridge at the fall 1988 equinox and discovered that the site is close to the east-west line passing through the center of the gap. As seen from 5AA8, the lowest portion of the gap, 2.2 miles to the east, has an azimuth of 89.6 degrees (see Figure 6.22).

FIGURE 6.23. Equinox sunrise viewed from the Overlook Pueblo.

Since the ridge offers many equally suitable locations for habitation, it seems likely that this particular location was chosen because of its alignment relative to the chimneys.

Because the gap between the chimneys has an elevation of 3.5 degrees and is 0.4 degrees away from due east, as seen from 5AA8, the sun will not rise between the chimneys on the morning of the equinox but will appear between the pinnacles on the morning of March 25. That date is close to half of the day count between summer solstice and winter solstice.

Summary

In the words of modern-day realtors, the benefits of the Chimney Rock area can be described as "location, location, and location." The irregular distant horizon provided a ready-made calendar, set dramatically off on the north with the double chimneys. The different behavior of the sun and moon could not be missed, especially when the rising moon was able to reach the gap between the chimneys. Construction of the Chimney Rock Pueblo on such an inaccessible location only makes sense if it was planned for ceremonies involving the moon. With an incredible view of the chimneys across the Piedra River, residents of the Overlook Pueblo could have watched sunrises near equinox from their doorways, rooftops, and plaza. At other times of the year, the sun rose between the chimneys as viewed from other smaller sites along the ridge. That information could have been passed directly over to the Chimney Rock Pueblo. With all of these opportunities for precise observations of the sun and moon, the Chimney Rock area may have functioned as the local "Greenwich Observatory" for Chaco Canyon. Runners could have carried calendrical information southward to the area around Salmon Ruin and then along the North Road into the canyon. Alternately, visual signals could have been sent southward to Huerfano Peak and then on to Pueblo Alto. The meaning and fate of the Chimney Rock community appear to have been closely tied to Chaco Canyon, and it was abandoned some time after 1130, in union with the dissolution of the Chaco regional system. The area was never reoccupied.

CHAPTER 7

YELLOW JACKET

oday, the lands north of Cortez are tranquil, dramatic, and fertile. The soil is plowed into rolling red swells. Pinto beans, wheat, and alfalfa provide striking contrasts of green upon furrowed red. In the quiet light of late afternoon when the furrows are darkened by long parallel shadows, it is a place of unforgettable beauty. The plowed fields surround scattered Ancestral Puebloan ruins, small broken castles of rocks overgrown by tall stands of sage. Above and beyond the fields to the east rise the mountains of the La Platas. To the south is the rounded mass of Sleeping Ute Mountain. At night the stars shine in the clear air as brightly as at any mountain observatory and continue undimmed down to the dark mountains of the east and south. The rainfall of 16 inches per year is twice that of either Hovenweep or Chaco Canyon, and it is easy to understand why the Ancestral Puebloans would have chosen such a place as this

---

The Astronomy of Yellow Jacket
Places to Visit (Contact the Archaeological Conservancy)

Solstice sunrise:	Sunrise from the solar monolith
Anytime:	Great kiva–Great Tower alignment to winter solstice
	Road connecting great house, great kiva, and solar monolith

in which to live. Today it is known as the Pinto Bean Capital of the World. In the 1100s it was probably the Bread Basket of the Chacoan World.

In this area known as the Great Sage Plain, there are the remains of eight major settlements, of which in the 1200s Yellow Jacket (5MT5) was the largest with a population between 850 and 1,360 (perhaps half that of Chaco Canyon).[1] It had a similar time line as Chaco; the neighboring Stevenson site (5MT-1) has been dated to 519 CE in Basketmaker III times, and was abandoned in the late 1200s.

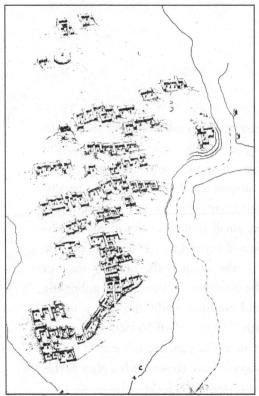

FIGURE 7.1. Yellow Jacket Pueblo.

Although the northern Ancestral Puebloans never reached the stage of building cities, the larger settlements, especially that of Yellow Jacket, may be viewed as incipient cosmic cities. With slightly larger populations and perhaps a more complex social structure, some of these larger Ancestral Puebloan settlements could have joined the ranks of classic ancient cities such as those of Mesoamerica: Teotihuacán, Copán, Palenque, and Tenochtitlán. Each of these major ancient cities in Mesoamerica was designed to be cosmic in scope, the mythical center and birthplace of the universe.[2] For those who lived in its vicinity, Yellow Jacket too may have been the center of the universe.

The cosmic cities of Mesoamerica appear to have functioned as marketplaces, fortresses, and administrative centers, as well as sites for ritual and ceremony. They were nourished both by pilgrims and by the inflow and outflow of foodstuffs and goods. As a result, they acquired great economic and political power.

Ceremonial centers were self-amplifying. They stimulated population increases, and their populations were organized in the construction of monumental structures, such as the pyramids of the Maya, the palaces of China, and the temples of India. The development of a high level of agricultural technology was clearly necessary. The leaders must have had considerable administrative sophistication: food surpluses had to be collected, stored, and redistributed; irrigation systems had to be designed and repaired; and agricultural land needed to be developed and maintained.

The organization of the ceremonial centers appears to have been inspired by a way of thinking that has been called "cosmo-magical," in which builders perceived a relationship between the celestial order above them and the biological rhythms of life. The two realms were seen as parallel in structure and synchronized in time. The cosmic city was aligned with the cosmos. Its streets and structures were often carefully oriented to the cardinal directions. The pattern of life within the city, in its festivals and celebrations, resonated with the movements of the sun, moon, planets, and stars. The goal, either intuitively felt or officially imposed, was a synchronization of individual human life with the larger universe. Such a goal of harmony between the individual and the larger cosmos is evident in the lives of the Pueblo peoples as well as in the structures built by their ancestors.

Yellow Jacket

Situated at an elevation of 6,800 feet, the Yellow Jacket ruin lies on a nearly flat mesa above Yellow Jacket Creek. With a sharp, serrated, eastern horizon, close to water and good farmland, it is convenient for living and superb for astronomical observation.

Yellow Jacket was one of the first of the northern Ancestral Puebloan ruins to be explored, having been visited by the geologist J. S. Newberry in 1859. The ruin is located near one of the major springs of the high plains between the Abajos of Utah and the San Juan Mountains of Colorado. The spring was a watering place and stop for stages in the late 1800s.

The first detailed survey of the site, designated 5MT-5, was made by Professor Joe Ben Wheat of the University of Colorado.[3] Professor Wheat has been the major figure associated with the archaeology of Yellow Jacket since 1954, when he began to excavate sites in the area. His first excavations were of a Basketmaker pithouse of the Stevenson site, 5MT-1. Following that, he has led many subsequent excavations yielding a uniquely comprehensive picture of the area. He considered the large ruin to be the major unexcavated Ancestral Puebloan site in the Southwest and believed it to have functioned primarily as a ceremonial center.

A Walking Tour

Unlike the ruins of Chaco Canyon, Mesa Verde, or Hovenweep, Yellow Jacket lies entirely upon private land. Permission to enter the site must be obtained from the owners, the Archaeological Conservancy and the Arthur Wilson family.

The main Yellow Jacket ruin (5MT-5) contains the highest concentration of kivas of any Ancestral Puebloan site in the Southwest (see Figure 7.2). Most of the kivas are organized in rows aligned approximately east-west, visible today only as shallow, saucer-like depressions, 15 to 25 feet in diameter, covered with thick, deep sage. Typically, north of each kiva

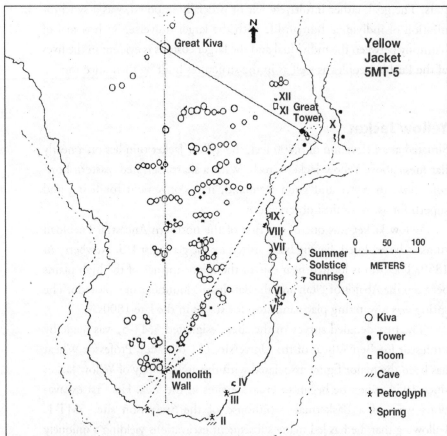

FIGURE 7.2. Map of the kivas, towers, and shrines of Yellow Jacket.

depression there are remnants of a block of rooms running parallel to the line of kivas. Next in abundance to the kivas and associated room blocks are towers, some twenty of them distributed among the kivas. In the northern edges of the ruin there is a large mound, perhaps the remains of a Chacoan great house.

I invite the reader to take an imaginary guided tour of this ruin. It will not be as fulfilling as the actual experience, but it has the advantages of avoiding the ear-loving gnats and the thick clumps of sage. We start at the solar monolith in the southern edge of the ruin. To the northeast are the three great snow peaks, Mt. Wilson, Wilson Peak, and El Diente, all over 14,000 feet. Directly east are the many summits of Mt. Hesperus and its companions, still mantled with snow in early June.

Rabbits scamper through the sage, and turquoise lizards sun themselves; there is even an occasional rattler. The sun appears to have been the major astronomical object that the Ancestral Puebloan observed, and the standing stone, 5 feet high and shaped into a wedge at its top (Figure 7.3), is the most obvious evidence of solar ceremonialism on the site. Its wedge-top is aligned toward Wilson Peak near the location of the sunrise on June 21. On the morning of the summer solstice, the pointed shadow of its top falls across the ruined wall of the room just to the west of the monolith. One summer we brought a piece of plywood, painted white, and set it up in front of that wall. Just at sunrise the top of the monolith cast a sharp pointed shadow on the wall (Figure 7.4). Throughout the weeks before the solstice, the position of the sharp top of the shadow could have been marked on the plastered wall, so that the day of solstice could have been easily anticipated.

To the west and east are four additional fallen monoliths of similar size (Figure 7.5). These monoliths and many smaller blocks define a wall directed toward the solstice sunrise point. The Ancestral Puebloans spoke to no living informant nor left us any written description of what they considered sacred and important in their lives. But here in this line of stones is a dimly heard voice and a faintly seen astronomical ritual. The distant mountains and the rising sun were clearly very important to those who lived here.

Beyond the monolith wall there is one break in the rubble, a narrow passageway just north of the standing monolith. Continuing past the monolith, we enter the great basin, surrounded on three sides by rooms,

FIGURE 7.3. Standing monolith.

FIGURE 7.4. Sunrise at summer solstice at the standing monolith. The shadow of the monolith is cast upon a board placed at the location of a suspected wall.

FIGURE 7.5. Eastern monolith.

kivas, and towers. On the western edge of the basin is a partially broken earthen wall that may have served as a dam, suggesting that perhaps during wetter times this great bowl functioned as a catchment basin. Alternatively, that wall may have served as a symbolic boundary, separating the outside world from the ceremonial space inside.

We move northward past the great basin along what is now a well-defined northern road, some 30 to 45 feet wide in places, passing between many pairs of kiva rows. Here we are in the heart of the ruin, in the kiva complex containing some eighty closely spaced, aligned kivas. After walking for more than 1,000 feet north of the solar monolith, we encounter the great kiva, a rounded crater 60 feet in diameter. The great kiva has been placed so carefully along the north-south line that it departs by only one-half of a degree from true north as seen from the solar monolith (Figure 7.12).

Farther to the north of the great kiva rises the northern mound, the likely remains of a Chacoan great house. A portion of the mound has been unfortunately carted away for road fill some time in the past. Standing on its top and looking southward, we begin to appreciate the size and complexity of this ruin. A short distance northward, plowed fields begin. The regularity of the swells of the land hints at even more kiva rows. How much farther north from us does the ruin extend? We may never know, for much has been lost under the farmers' plows.

Solstice Sunrise

We were drawn to Yellow Jacket by the mystery of solstice sunrise, and in the coolness of dawn we rose on many mornings during our summer fieldwork to watch the brightening sky above the eastern mountains. Each day before the summer solstice, the sun rose slightly northward of the previous morning. Its northward movement slowed noticeably as it approached the great massif of Wilson Peak. The rising sun climbed the southern side of the massif, and then one morning it stopped and paused. The mountain appears to act as a barrier to the northward-moving sun. Did the sun seem to the ancient sun priest to be returning to his northern home after a year's journey (see Figure 7.6)?

For the Hopi, the sun may be an animate being who travels back and forth from south to north, from his southern house to his northern house.

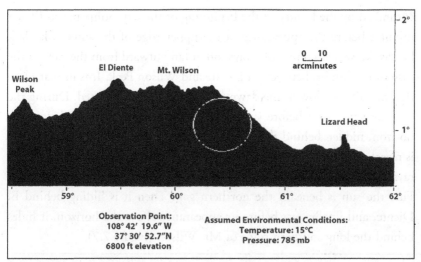

FIGURE 7.6. Computer simulation of sunrise on summer solstice.

Offerings are made by the Hopi when the sun is in his summer and winter houses, requesting his benevolence toward them and their crops. As seen from Yellow Jacket, such a northern house would have been the Wilson–El Diente massif. Not merely an abstract point on the horizon, it may have been viewed as a real habitation in which the sun resided unmoving for some four days.

Today, somewhat wiser by nearly 1,000 years of astronomical observation and theory, we approach the dawn with a different set of preconceptions. The sun rises each morning because the earth turns on its axis and in that slow turning, a hot ball of gas is carried into view. But it is easy to imagine an ancient sun watcher, standing at the edge of the mesa in the dim light, waiting for the first evidence of the sun in his summer house,

FIGURE 7.7. Solstice dawn at Yellow Jacket, crepuscular rays over Mt. Wilson.

enthralled by the beauty of the beginning of the day. Some ten to fifteen minutes before the appearance of the upper edge of the sun's disk, dark beams, known as crepuscular rays, often fan outward from the cleft in the eastern mountains between El Diente and Wilson Peak. It is in that place that the diffuse glow of the dawn first becomes concentrated. During the minutes remaining before sunrise, the sun creeps southward below the horizon, hidden behind the Wilson–El Diente massif. The sun's presence is revealed by the increasingly bright glow of dawn and by the crepuscular rays, which rotate slowly counterclockwise while their focus moves south. First the sun is beneath the northern gap. Then it is hiding behind El Diente, and finally, just before its appearance above the horizon, it hides behind the long southern ridge of Mt. Wilson (Figure 7.7).

FIGURE 7.8. Solstice sunrise from the Yellow Jacket monolith.

When the sun eventually appears above the southern ridge of Mt. Wilson, it is blinding, red, and gorgeous, but it is no surprise (Figure 7.8). Its rising point has been fully anticipated. That first hint of concentrated brightening is a magic time when the uncertainty of night is replaced by the certainty of the sun's location, when creation of the day seems assured. We can now only speculate about the complex meaning of those stones of the solstice line and the great effort that went into their careful placement. They speak to us of the sacredness and mystery of those eastern mountains and the solstice sunrise.

Anticipation of the Summer Solstice

As seen from the ruin, there is a small horizon feature just to the south of Mt. Wilson, with a flat top and straight sides that resemble the rock monoliths themselves. The monolith-like feature, especially dramatic and

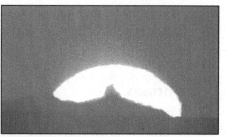

FIGURE 7.9. Yellow Jacket sunrise on June 5 behind Lizard Head.

prominent during the half hour just before sunrise, may be another example of anticipation in the life of the Ancestral Puebloans. The horizon rock is Lizard Head Peak, 13,113 feet in elevation, and on the morning of June 5 the sun rises behind it as seen from the southern portion of the ruin (Figure 7.9). This unusual event provides a precise and dramatic observation by which to anticipate the date of the summer solstice.[4] After observing the first sunrise behind Lizard Head, the sun priest would know that some sixteen days remained until the actual solstice. By notching a calendar stick or tying knots in a rope, he could accurately determine the number of days remaining as the sun continued to move farther north toward Mt. Wilson. Because the sun has no noticeable motion along the horizon at the time of solstice, accurate determination of solstice requires observations at earlier or later times. Furthermore, sufficient time for preparation for the ceremony was probably needed: dances had to be rehearsed, costumes made, and food prepared. Similar opportunities for anticipation of the solstice are found in Chaco Canyon at Piedra del Sol and Wijiji.

Whatever meaning the sunrise behind Lizard Head may have had for the Ancestral Puebloans, it is, even today, a remarkable event. As the red sun struggles to rise above the horizon, it appears to be stabbed by the rock peak. We wonder what stories may have grown up around that astronomical event. It is a long and perilous journey for the sun, a journey vitally necessary for the earth to break out of the winter's cold. As an animate being, the sun may have been subject to the vicissitudes and uncertainties of the natural world. Mesoamerican deities—Quetzalcoatl, for example—were often fragile beings who could suffer pain and defeat. Quetzalcoatl lived in uncertain relationships with the unpredictable natural world. In Mesoamerica as well as in a number of other cultures, the daily and annual cycles of the sun are depicted in terms of battles with the forces of darkness and disorder. The sun's encounter with Lizard Head, as seen from

Yellow Jacket, may have been viewed in local myth as one of those challenges and tests that the sun had to undergo each year.

Shrines

Located primarily along the eastern half of the perimeter of 5MT-5 are twelve places that may have been prehistoric shrines. In the most general sense, a shrine is a human-made structure or a natural feature held in sacred esteem. Pueblo shrines are often located on the tops of mesas, are typically to the east of habitation sites, are frequently stone enclosures, and may be associated with rock art. Pueblo shrines are often locations where offerings are placed and can be features such as caves, mountains, rock formations, springs, and pools.

The great Ancestral Puebloan migration of 1300 CE resulted in the movement of people living north of the San Juan River into the Rio Grande valley and the Pajarito Plateau of northern New Mexico. It is to this area and its ancient and modern inhabitants that we turn for clues about the meaning of shrines. A well-known living shrine on the Pajarito Plateau is that of the Stone Lions in Bandelier National Monument. Within a circular ring some 20 feet in diameter, enclosed by large slabs of volcanic tuff, lie two crouching lions facing southeast. Each lion is about 6 feet long, including 2 feet of tail. The shrine is now visited by Indians from several pueblos and is apparently dedicated to hunting. Visitors who walk from the monument headquarters in Frijoles Canyon to the Stone Lions find broken pottery, shells, feathers, petrified wood, obsidian, many bleached antlers, skulls, and bones of deer and elk. The shrine is about a half a mile northwest of the ruins of the village of Yapashe, the traditional ancient home of the Cochiti.

According to modern Puebloan traditions in the Rio Grande valley, shrines on high points protect the world.[5] They are especially powerful when they are placed on mountains at the four cardinal directions of space. Around the Pajarito Plateau are stone enclosures that may have been shrines on high places and hilltops near almost every important ruin, most often to the east of each ruin.

Caves are important shrines for the modern Puebloans, who believe that they can come into close contact with the powers of the underworld when they store ritual implements and make offerings in caves.[6] Earth Mother remained underground after the first people emerged through the sipapu. Specific caves and small lakes are considered by various Puebloans as the sipapu opening. All springs are assumed to connect with the underground lake from which come the rain spirits. Offerings presented to the Earth Mother may be ears of corn with prayer feathers attached, feather bunches, prayer sticks, food images, or small pieces of food, and they may be placed in caves or substitute sipapu shrines such as circular stone enclosures.

The cult of caves is one of the most ancient of the pan-Mesoamerican world. Caves are frequent features of the Mesoamerican sacred landscape and are pictured extensively in rock paintings, murals, and codices going back to the time of the founding of Teotihuacán.[7] The family of symbols, including caves, water, mountains, fertility, and maize, form the nucleus of Mesoamerican religion. The rain and mountain god, Tlaloc, corresponding to the Mayan god Chac, was one of the main deities of ancient Teotihuacán. Together with the sun and war god, Huitzilopochtli, he presided over the fierce and bloody rituals performed on the Templo Mayor in Tenochtitlán.[8] These symbols of water and sun, placed upon the summit of the great pyramid-mountain of Templo Mayor, dominated the Aztec world.

Sacred springs and pools are also located at the cardinal points of several of the Rio Grande pueblos. Until 1902, the major ceremonialists of Acoma would make a semiannual trek with solar offerings to their sipapu lake or spring, somewhere in the vicinity of Cortez.

The largest stone enclosure on the mesa top at the main Yellow Jacket ruin is a semicircular enclosure to the south and east of the solar monolith (Figure 7.10). The enclosure is open precisely to the east, has a diameter of nearly 20 feet, and is constructed out of large sandstone slabs averaging 2 feet in length. This structure is similar in size to the circle of stones surrounding the shrine of the Stone Lions in Frijoles.

A short distance to the north is a similar but smaller enclosure, also open to the east. It is aligned with the monolith wall and was placed just above the largest cave in the eastern cliffs. The enclosure was constructed on the line between the standing monolith and the position of the rising sun. Its location is similar to that of a possible shrine near Shabik'eshchee in Chaco Canyon, which is located between a sun symbol and the position of sunrise on summer solstice.[9]

The location of this enclosure in the direction of summer solstice suggests that it functioned as a site for offerings to the sun. Immediately

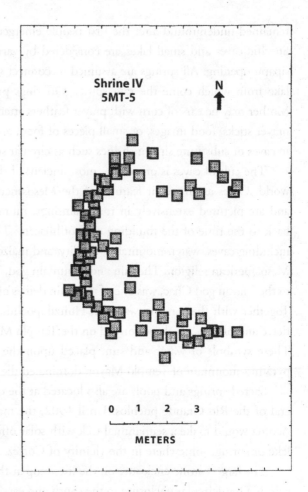

Shrine IV
5MT-5

N

0   1   2   3
METERS

FIGURE 7.10. (top) Southeastern shrine, triangulation of stones; (bottom) drawing of southeastern shrine.

below is the large eastern cave, which, if dedicated to the Earth Mother, would have provided a powerful conjunction of sun and earth symbolism. The eastern cave has an opening more than 60 feet across and is 12 feet tall at its highest. Numerous shaped stones across its front indicate that buildings had once been constructed within the cave. A tower, kiva, and small block of rooms are located on the slope below the cave. Farther down the slope is a series of terraces, which because of their sheltered position may have been used for early planting of ceremonial crops.

Great and Small Kivas

The four mapping projects at Yellow Jacket have identified slightly different numbers of kiva depressions. Our mapping project in 1987–1988 included only those kiva depressions for which we could identify a clearly defined center and rim (Figure 7.2).[10] Generally, when we surveyed a kiva depression with the theodolite, we measured five points on the rim and the center. Our work located the position of 117 kivas, of which 79 were contained in east-west kiva rows. North of the kiva rows are blocks of rooms, and to the south are middens. What is probably the definitive survey of the site has been performed by Crow Canyon; their map shows some 180 kivas.

The ceremonial heart of Yellow Jacket may have been the great kiva at the northern end of the north-south road. Totally unexcavated, it now is an impressive bowl some 69 feet in diameter, set in a forest of dense sage (see Figure 7.11). The center of the intermediate kiva, 600 feet to the south, lies only slightly more than an inch off true south from the center of the great kiva. As seen from the center of the intermediate kiva, the center of the great kiva would have been slightly less than 1 minute of arc away from true north. The smallest angle that the unaided human eye can discern is approximately 1 minute. The great kiva may have been linked to the intermediate kiva by a ceremonial pathway designed with an extraordinarily accurate alignment.

The intermediate kiva also lies due west of the Great Tower. From the center of the intermediate kiva, the Great Tower has an azimuth of 90 degrees, 58 minutes of arc. The line running due east from the center of the kiva crosses the rim of the Great Tower. An observer within the Great Tower, looking through an opening in its walls, may have been able to

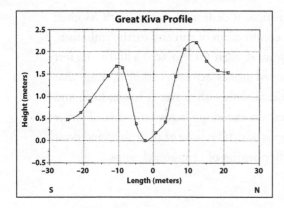

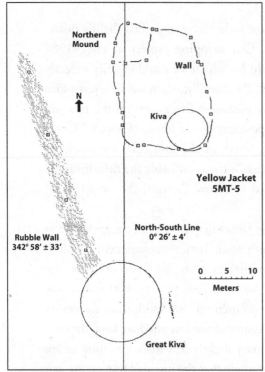

watch the sun at the equinox setting directly over the intermediate kiva or the sun at the summer solstice setting over the great kiva. I'm afraid we shall never know.

Lines to Ancient Skies

Many of the structures of 5MT-5 lie along three prominent lines, each of which points to a significant direction in the heavens (Figure 7.2). The three lines served no obvious utilitarian or architectural purposes, other than delineation of space or astronomical symbolism and ceremony. They radiate outward from the solar monolith with a spacing of approximately 30 degrees.

The great kiva, the intermediate kiva, the northern mound, and the standing monolith are members of the north-south group, which has an azimuth of only 26 minutes of arc east of true north and a length of over 1,600 feet (see Figure 7.12). This line departs from true north by less than the angular diameter

FIGURE 7.11. (top) Great Kiva, north-south profile; (bottom) North-south line is shown passing through the great kiva and the northern mound.

of the sun. The accuracy of this alignment is comparable to that found at Casa Rinconada in Chaco Canyon. By comparison, the accuracy of placement of these structures still cannot compete with that of the greatest monument of the ancient world, the pyramid of Kufu at Giza, erected about 2600 BCE. The worst alignment of its huge base, covering more than 13 acres, is on the east side. It departs from true north by a mere 5.5 minutes.

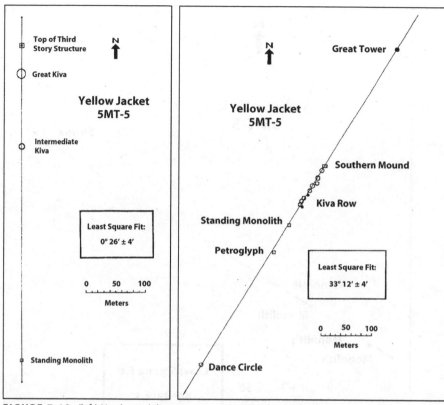

FIGURE 7.12. (left) North-south linear group connecting the northern mound and the standing monolith.
FIGURE 7.13. (right) 33-degree line connecting the Great Tower of 5MT-5 and the Dance Circle of 5MT-3 on the next mesa to the southwest.

The intermediate kiva emerges as an important and unique structure, with the great kiva to the north, the Great Tower to the east, and the standing monolith to the south. Unlike most of the structures of 5MT-5, the great kiva, the intermediate kiva, and the Great Tower lack associated middens, indicating that they were never used for extensive habitation.

The road southward from the great kiva of 5MT-5 may have been the locus of a ceremonial journey similar to those that have been proposed for the road systems of Chaco Canyon. Celebrants may have left the great kiva, symbolic of the place of emergence, moved past the kiva rows and through the basin of sacred water, toward the symbol of death, the Cave of Ashes.

Slightly more than 30 degrees from true north, the second line connects the Great Tower of 5MT-5 with the Dance Circle of 5MT-3 on the next mesa to the southwest over a distance of some 2,500 feet (Figure 7.13).

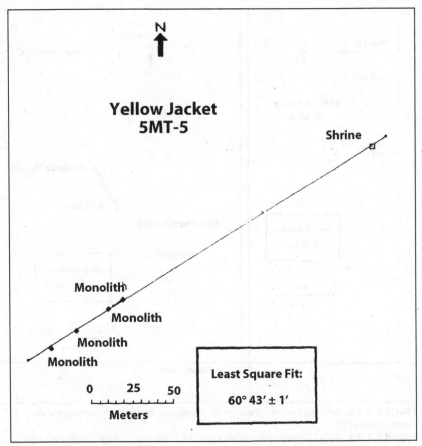

FIGURE 7.14. Group consisting of monoliths and possible shrine aligned toward the position of the sunrise at summer solstice.

The line includes other important features, such as the southern mound, a row of eight kivas (the longest kiva row on the site that is not aligned approximately east-west), two towers, the standing monolith, and the southwestern petroglyph. The perpendicular to the line lies within a few degrees of the winter solstice sunrise.

The third line, which oriented with the summer solstice sunrise, is rotated approximately 60 degrees away from true north (Figure 7.14). Its length, from the fallen western monolith to the eastern stone enclosure at the edge of the cliffs, is nearly 760 feet. It too may have been a ceremonial pathway leading from the high point near the standing monolith to the possible shrine above the eastern cave.

The solstice line also serves as a symbolic boundary. Southeast of the solstice line, the only structure on the mesa top is the large enclosure open to the morning sun. Symbols of darkness, birth, and water are present to the north of the line. To the southeast of the line, the sun and the symbols of fire and death dominate.

These three lines serve to join together important artificial and natural features of the landscape. Each starts in the north with major symbolic features: (1) the monumental structure of the great kiva, perhaps representing the origin of the world; (2) the northeastern Great Tower; and (3) the mountains, the eastern cave, and the solstice sun.

And there are more lines. A fourth line, deviating only slightly from true east, proceeds from the southern mound to the eastern cave. The kiva rows and their middens also run approximately east-west. The fifth line, from the great kiva to the Great Tower, leads toward the position of the sun at winter solstice. Each of these lines continues to be separated by approximately 30 degrees. The sixth line, which should have an azimuth of 150 degrees, has not yet been found.

The importance of carefully defined linearity to the builders of 5MT-5 is strongly implied by the results of our surveys. Elsewhere in the Ancestral Puebloan world, ceremonial linearity appears to be present in the extensive system of roads in Chaco Canyon. The northern terminus of the North Road is perhaps symbolic of the origin of the world, and passage along the road may have involved reenactment of a creation mythoogy in which the first people emerged from the underworld and moved southwest to their present homes.

The Development of Yellow Jacket

There are four phases to the development of Yellow Jacket that we can identify:[11]

1. People first settled in the Yellow Jacket area as early as 519 CE, based upon excavations of pit houses by Joe Ben Wheat at the Stevenson site, 5MT1, slightly to the west of the main ruin.

2. Chacoan people or strong Chacoan influences appeared at 5MT-5 around 1080 CE, when the great house and great kiva at the northern part of the site were built. The solar monolith, some 1,700 feet south of the great house, as well as the east-west-aligned kiva rows, may date from this period.

3. The major surge in population occurred after the decline of Chacoan influence around 1180 CE. During this period the population may have grown to ten times that of the earlier period. The new kiva rows departed from true east-west by an average of 5.4 degrees and may have been built along a new axis of the site running from the great kiva to the Cave of Ashes on the southern edge of the peninsula.

4. Finally, during the period 1240–1280 CE, towers were built near the spring at the northeastern corner of the site. Close by is the Great Tower, which is larger than any tower found in the Hovenweep area. A tree ring date of 1254 CE may establish its date of construction. There are sixteen other towers at 5MT-5, of which ten are close to kivas to which they may have been connected by tunnels. These tower-kiva pairs may be ceremonial, symbolically combining the earth and the heavens, or they may have been defensive, providing escape routes from the kivas in times of danger.

Conclusion

Whatever the meaning of its architectural symbolism, the complexity of organization of 5MT-5 is an important revelation about the sophistication of Ancestral Puebloan society. The obvious care with which the linear groups have been established, the intricate interrelationships of the groups, and the manner by which the groups integrate features of the natural environment indicate the importance of astronomical ceremony for the designers and inhabitants of Yellow Jacket.

The Ancestral Puebloans paid attention to the heavens, but they did more. The picture that emerges is that of an astronomical infrastructure, consisting of people with skill, knowledge, and power. Certain individuals had the techniques for determining accurate cardinal directions, the knowledge of the short and long cycles of sun and moon, and the political authority to influence major construction projects. There must also have been a persuasive belief system containing astronomical motifs that provided justification for such major endeavors as the construction of the Chimney Rock Pueblo and the alignment of the great kivas.

Furthermore, we appear to be encountering a highly tuned sense of the harmony of nature. Rituals expressive of such harmony apparently could not be casual affairs but had to be performed with accurate geographic orientation in order to be effective. Both cosmology and cosmogony must have played roles. Accurate cosmology is manifest in the attention to the alignment of human lives to the structure of the cosmos. Reenactments of the origins of the world may have involved rituals performed on the north-south roads of Chaco and Yellow Jacket and in the great kivas. Practical calendrics, in which the solar and lunar calendars were reconciled, may be manifest in the concern for the midwinter full moon rise at Chimney Rock, the corner windows at Pueblo Bonito, and the sun room at Hovenweep Castle.

Moving throughout all of these activities are the shadowy figures of skilled practitioners, who perhaps traveled and traded their astronomical skills throughout the Ancestral Puebloan culture area. In the thirteenth century, their roles may have intensified or radically changed. Their skills may have been desperately sought or their reputations denigrated when the harmony between the earth and sun faltered and the land began to fail as a provider.

CHAPTER 8

MESA VERDE

Few visitors to Mesa Verde National Park miss the opportunity to explore the Sun Temple, and few can leave that extraordinary structure without a mixture of admiration for its builders and puzzlement over its meaning. With its careful design and symmetry, it is clearly not a purposeless or unplanned building. In fact, it is the largest exclusively ceremonial structure built by the Ancestral Puebloans.

The Astronomy of Mesa Verde
Places to Visit

Solstice sunrise:	Sunrise over the La Platas from Far View House
Anytime:	Sun Temple: sun rock, pecked basin, solar and lunar alignments to Cliff Palace.
	Cliff Palace: pictographs in four-story square tower: 75 tick marks; rug-like pattern; horizon profile of peaks.
	Cedar Tree Tower: pecked basin; tunnel between kiva and tower.

When he excavated the building in 1915, archaeologist Jessie Fewkes concluded that it must have been primarily dedicated to ceremonial activities.[1] The effort expended in pecking and shaping nearly every stone of the wall veneer indicated the importance of the building for the people

FIGURE 8.1. The Sun Temple.

who lived in its vicinity. Built on a narrow peninsula, the Sun Temple is bordered by Cliff Canyon to the north and Fewkes Canyon to the south. In the Cliff and Fewkes Canyons area are some thirty-three habitation sites, including Cliff Palace, Oak Tree House, Fire Temple, Mummy House, and Sunset House.[2] With a total of some 530 rooms and sixty kivas, the population in the vicinity of the Sun Temple may have exceeded 600.

Cliff Palace faces a southwestern horizon where the sun sets on December 21, the evening of winter solstice. The horizon is a nearly flat expanse devoid of major markers, except for the Sun Temple on the opposite mesa, at a distance of 945 feet across Cliff Canyon. Is there a possibility that that the exclusively ceremonial structure of the Sun Temple was connected to Cliff Palace through astronomical ritual? In the summer of 1991 Frank Occhipinti, a graduate student in anthropology, and I measured many sites in Mesa Verde, testing them for possible astronomical alignments. On June 7 we set our theodolite upon the northeast wall of the Sun Temple, aligned it to the calculated position of sunset at winter solstice, and then "flipped" the telescope to look backward at Cliff Palace. We found that all of the main structures were 5 to 10 degrees to the north. The only place where someone could have watched sunset over the Sun Temple was at the far southern end of the Cliff Palace enclosure, near the exit ladders leading to the rim of the canyon. Frank remembered seeing a

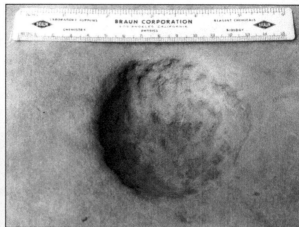

FIGURE 8.2. Pecked basin in Cliff Palace.

platform in the area, and we went over to investigate. Not only did we locate the platform, but Frank found a pecked basin in its center (Figure 8.2). The basin has a diameter of 3.2 inches and a depth of 1.2 inches, which might have been a place for a sun priest to stand to watch sunset. We found that the center of the Sun Temple (halfway between its two circular rooms) as viewed from the platform has an azimuth of 235 degrees, 24 minutes of arc, which is within 6 minutes of arc of the position where, according to our calculations, the lower limb of the winter solstice sun would have touched the top edge of the perimeter wall of the Sun Temple (Figure 8.3).[3] We subsequently were able to verify the alignment at December solstice (Figure 8.4).

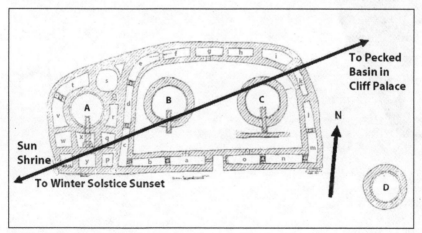

FIGURE 8.3. Sun Temple, alignment to winter solstice sunset. Note the sun shrine at the southwestern corner.

FIGURE 8.4. December solstice sunset over the Sun Temple photographed near the southern pecked basin in Cliff Palace.

FIGURE 8.5. Sun shrine at the southwest corner of Sun Temple.

Among the historic Puebloans there has been a distinction between offertory shrines to the sun and calendrical sun-watching stations.[4] The pecked basin may have marked a sun-watching station.[5] The strangely eroded rock attached to the southwestern corner of the Sun Temple (in the lower left of Figure 8.3; Figure 8.5) may have been an offertory shrine. The offertory shrines were often associated with unusual stones, concretions, or oddly shaped rocks such as the one contained in the stone enclosure. Such shrines were sites, usually some distance from the pueblo, where water, cornmeal, prayer sticks, or prayer feathers were placed. The natural features of the stone look like the rays of the sun, and it was this object that led Fewkes to suggest the structure was a sun temple. This stone may have been the original solar feature of the mesa and used for offerings to the sun well before the construction of the Sun Temple.

Down in Fewkes Canyon, just south of the Sun Temple is another fascinating structure, called the Fire Temple by Fewkes. It contains a large circular fire pit with a diameter of nearly 6 feet. Fewkes speculated that it may have been involved with solar related fire ceremonials similar to those of the Hopi. In the Fire Temple, the eastern horizon is partially blocked by the mesa cliffs below the Sun Temple. Those cliffs block the view of the rising sun until mid winter, when a shaft of sunlight near dawn falls on the fire pit (Figure 8.6). A fascinating possibility is that the winter solstice celebration in the Cliff and Fewkes Canyons included the lighting of a fire

FIGURE 8.6. Light of the winter solstice sun striking the fire pit of the Fire Temple.

in the fire pit at the moment it was illuminated by light of the morning sun. That celebration may then have been concluded when the sun set over the Sun Temple as seen from Cliff Palace.

Back in 1991, there was one remaining alignment at the Sun Temple to be investigated. Early in the morning of June 30, I set up my theodolite on its perimeter wall so that it was exactly on the line tangent to the outer walls of the two circular rooms of the Sun Temple. I flipped the telescope of the theodolite again, and this time I found the cross hairs center on the doorway of the square tower of Cliff Palace (Figure 8.7). If the two circular rooms of the Sun Temple had been towers extending above its perimeter wall, they could have functioned as a "gun sight" for an observer in or near the four-story tower. The line of sight from the center of the T-shaped doorway in the fourth story through the gap formed by the double towers has an azimuth of 227 degrees, 2 minutes. The sun never reaches that far south, but the moon at major southern standstill would have touched the perimeter wall of the Sun Temple at an azimuth of 227 degrees, 9 minutes, and thus would have fitted nicely in the gap between the towers every 18.6 years (Figures 8.2 and 8.3).

The intersection of the lunar standstill line of sight with the square tower is particularly interesting because pictographs painted on its interior

FIGURE 8.7. Cliff Palace. Visitors have lined up to look into the doorway of the Square Tower.

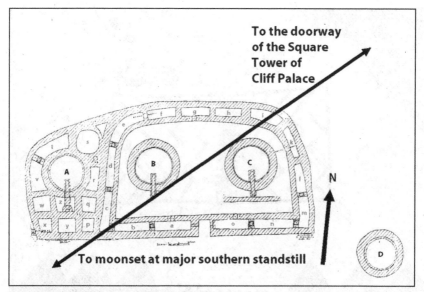

FIGURE 8.8. Sun Temple, alignment to major lunar southern standstill.

walls, some of the finest pictographs in the park, may be associated with the moon. On the interior wall of the third story of the tower there are four vertical lines, each containing seventeen to twenty tick marks, appearing more like tally marks of events than decorative designs (Figure 8.9). The total number of such marks is seventy-four to seventy-five, corresponding to an average of 18.5–18.75; these marks may be records of four lunar standstill cycles observed from Cliff Palace. Tree ring dates in the Cliff Canyon and Fewkes Canyons area span the period between 1180 and 1279. The exact dates of initial habitation and final abandonment are unknown, but four major standstill cycles could have been observed at 18.6-year intervals from Cliff Palace, starting with the standstill of 1186 and continuing through the standstill of 1261.

A second pictograph that may also have a lunar association is found at approximately the same height as the four lines (Figure 8.10). Contained within a rectangular border, the figure is divided by a vertical line with approximately twelve marks, and on either side there are twelve zigzags. Such a pattern is not uncommon in Ancestral Puebloan art, but the recurrence of twelve marks and twelve zigzags is noteworthy because of the twelve "moons" during the year. In one month, the moon swings from southern extreme to northern extreme and back to southern extreme. That pictograph may be a representation of the changing positions of the rising

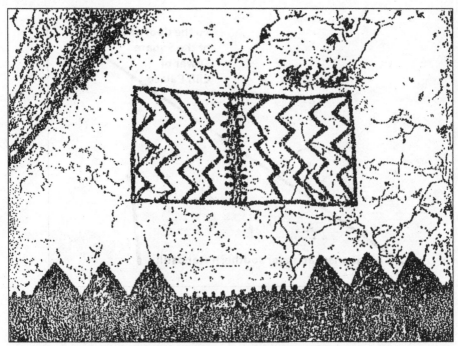

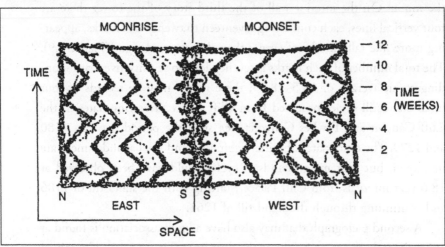

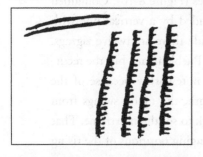

FIGURE 8.9. (left) Four sets of tick marks in the four-story tower of Cliff Palace.

FIGURE 8.10. (top and middle) Pictographs in the four-story tower of Cliff Palace, possibly representing the varying positions of the moon on the horizon.

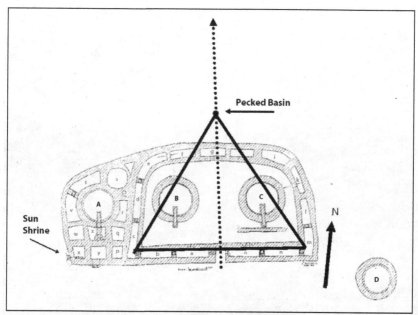

FIGURE 8.11. Sun Temple, alignment to Cedar Tree Tower.

and setting moons during a one-year period. It is the kind of diagram that an astronomer of today might draw on a blackboard to illustrate the changing positions of the moon or an Ancestral Puebloan astronomer might have drawn for an apprentice.

Immediately beneath the rectangle is a third pictograph consisting of two sets of three triangles separated by twelve circles (Figure 8.10). The triangles may represent the La Plata Peaks on the northeastern horizon, and the circles may represent an annual series of sunrises or moonrises.

We cannot know for certain what was in the mind of the artist or artists who painted these pictographs. But the correspondences are impressive: they are located at the third floor of the tower where an observer could view the setting moon over the Sun Temple; they contain numbers corresponding to the lunar cycles of twelve months and 18.6 years; and, finally, the observing location in the tower is on the tangent line of the circular rooms of the Sun Temple, which intersects the setting moon at its major southern standstill.

The experience of standing at Cliff Palace to watch the moon set over the Sun Temple, perhaps between the double towers at the Sun Temple, is reminiscent of the similar situation at Chimney Rock. Could it be that the inhabitants of Cliff Palace were intentionally duplicating the experience of

FIGURE 8.12. Cedar Tree Tower, note the tunnel connecting the tower and kiva.

the moon rising between the double pinnacles of Chimney Rock? Chimney Rock was abandoned about 1130, and the moon may have been observed from Cliff Palace some fifty years later.

Around the time of a major standstill, the moon would have set behind the Sun Temple in a dark sky only during the spring in the six-month period between winter solstice and summer solstice. The moon would have first set over the temple after winter solstice as a slender crescent. Month by month, with impressive regularity, the setting moon would have grown in size, until the full moon would have set at sunrise near the day of summer solstice.

The rotation of the east-west axis of the Sun Temple by 10 degrees 40 minutes of arc to the west of true north may also have an astronomical explanation. The architectural precision of the Sun Temple suggests that a departure from true east-west may not have been accidental or arbitrary. A clue to this puzzle may be provided by a second, large pecked basin 16.4 feet north of the Sun Temple.[6] The basin lies on the perpendicular drawn from the southern wall and may indicate an important direction to the north. Standing above the alcove in the southern wall, looking across the center of the Sun Temple and across the basin beyond, one faces the direction of Cedar Tree Tower, some 1.24 miles to the north. The tower, 9 degrees 34 minutes west of north, is built of double-coursed walls in which the stones have been carefully pecked to fit the curvature of the walls. It is similar in the quality of its construction to the Sun Temple, and contains a third pecked basin in its bedrock floor that may mark a second sun-watching station.

That oval basin, (4 x 5 inches across and 2.3 inches deep) was described by both Fewkes and Rohn as a non-kiva sipapu, but there may be an alternate explanation. The feature has the pecking, depth, and diameter of the other pecked basins that we have found and may be part of a ceremonial network.[7] We have located some 200 pecked basins in Mesa Verde. Sipapus can be seen in the floors of about half of the kivas in Mesa Verde and are not visually interconnected. They typically are deeper and more cylindrical in shape.

There is a tunnel connecting the tower to the kiva just to the south. Another kiva-tower combination connected by a tunnel is found at Badger House. These kiva-tower tunnels may have been part of shamanic rituals connecting the earth and sky. Alternately, they may have been defensive in nature, providing an escape route from the kiva in case of an attack. The axis of the kiva has been carefully aligned to within 3 minutes of arc from true north-south. There is no midden or other evidence of habitation in the immediate vicinity of Cedar Tree Tower, suggesting that this place was primarily ceremonial. Some distance to the east is Painted Kiva House, which is the major residential structure in the area.

The northeastern horizon, visible from the tower, contains the prominent, sharp peaks of the La Platas, which would have provided excellent calendrical markings for the summer solstice sunrise. Announcement of the dates of ceremonies determined from observations of the sun on the horizon may have been visually communicated from the Cedar Tree Tower southward to the Sun Temple and Cliff Palace, and also northward to the Far View community, where there are other basins at the ends of sightlines.

Summary

Astronomical ceremonies associated with both the sun and the moon may have occurred at the Sun Temple and Cliff Palace. The orientation of the Sun Temple to the Cedar Tree Tower may have had an important ritual significance if a sun priest in Cedar Tree Tower could have communicated the date of summer solstice to the Sun Temple. Participants in the June ceremony at Sun Temple could have watched the sun rise above Cliff Palace. At the times of major standstills, participants in Cliff Palace would have watched the full moon set over the Sun Temple at the same time the sun was rising. At winter solstice, the rising sun could have been celebrated in the Fire Temple, and at the end of the day the setting of the sun over the Sun Temple could have been celebrated in Cliff Palace. At these ceremonies, participants may have been impressed by such demonstrations of the order of the heavens and of the power of individuals to identify and predict that order.

CHAPTER 9

THE TOWERS AND CASTLES OF HOVENWEEP

FIGURE 9.1. Square Tower.

The land near Hovenweep National Monument in southeastern Utah contains lonely towers perched on the rims of canyons, standing guard over scarce water resources. They are beautiful structures, built of carefully shaped and pecked sandstone. The last half of the 1200s was a desperate time, and the many towers built throughout the area may have been intended to protect the most vulnerable during an attack and to guard scarce sources of water. One can feel an almost palpable fear in these towers. The combination of the "Great Drought" from 1276–1299 and

The Astronomy of Hovenweep, Goodman Point, and Sand Canyon
Places to Visit

Solstice sunrise: Setting sun in portholes of Hovenweep Castle
Rising sun at Holly House
Unit-Type House: sunrise through porthole
Cajon Group: solstice light through porthole

Anytime: Sand Canyon in the Canyon of the Ancients
Goodman Point

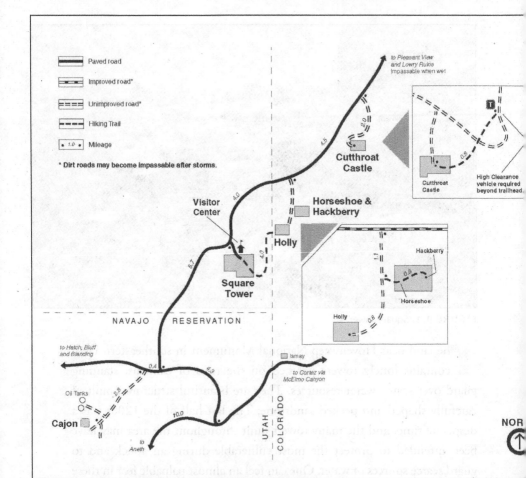

FIGURE 9.2. Major sites of Hovenweep National Monument.

the cooler temperatures at the onset of the Little Ice Age made dry-land farming increasingly unpredictable, if not impossible. During this period conflict reached unparalleled levels.[1]

Many of the isolated towers and structures of Hovenweep are difficult to enter and rely on small "portholes" to provide minimal light and ventilation. In spite of the threats they faced, or because of them, the residents of Hovenweep built astronomy into their villages. Shamanism and solar rituals may have gained power because people had little else to turn to during these times of drought and danger.

Hovenweep Castle

Named for its resemblance to the fortresses that dotted the European landscape during the Middle Ages, Hovenweep Castle stands on the rim of the main canyon, above a cool spring.[2] The D-shaped tower to which a number of rooms were attached is actually just a small part of the entire building as it once appeared. A rectangular room, dubbed the "sun room," was attached to the southern side of the tower in 1277, and it is this room's ground floor to which astronomers have been drawn. The late date of construction of the sun room is remarkable, at a time when life was becoming increasingly precarious.

FIGURE 9.3. Hovenweep Castle, note the defensive nature of the building's only doorway.

FIGURE 9.4. Hovenweep Castle from the south. Three portholes in the first story are visible; the one on the right is the port for June solstice sunset.

From this room, the Ancestral Puebloans of Hovenweep Castle could mark the equinoxes and both solstices. At the summer solstice sunset, a ray of light streams through a porthole in the sun room and touches the edge of the doorway into the eastern room. For only a few days around summer solstice, just as the light of the setting sun is fading, a spot of faint reddish sunlight just enters that room (Figure 9.4). The long north wall of the room, along which the light beam travels as the solstice approaches, may have been plastered and scored with vertical marks indicating the days before solstice festivals.

At winter solstice sunset, a porthole on another wall lets light fall on the lintel of the passage to the tower, and near equinox light enters through the doorway. These complex effects, first discovered by Ray Williamson and his colleagues, together with the alignments of Casa Rinconada and Pueblo Bonito, were some of the first indications of the careful attention to astronomical detail by the Ancestral

FIGURE 9.5. "Sun Room" of Hovenweep Castle showing sightlines to summer and winter solstice and the equinox.

Summer Solstice

Winter Solstice

Equinox

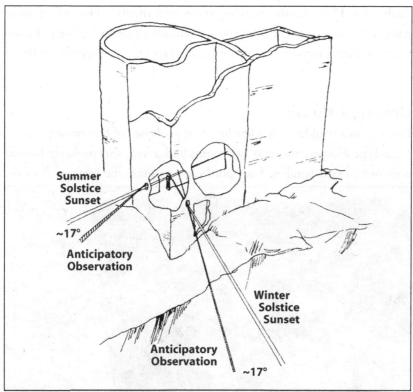

FIGURE 9.6. Diagram of Hovenweep Castle showing summer and winter solstice alignments. The possibility of anticipating the events is indicated. Note the similarity of the alignments of ports and doorways at winter and summer solstice.

FIGURE 9.7. Hovenweep Castle showing beam of light from setting sun at summer solstice. The two photographs show the movement of the patch of light just before sunset. Just as the light fades, it touches the edge of the doorway. Note that entry inside the building is now not permitted.

Puebloans.[3] However, the meaning of the solar rituals at Hovenweep must have been vastly different from those ceremonies that took place in Chaco Canyon during those peaceful and secure days of the eleventh century.

Unit-Type House

Perched on a boulder only a few hundred yards east of Hovenweep Castle, Unit-Type House consists of six rooms and a kiva and probably housed only one or two families. Four portholes perforate the intact wall of the

FIGURE 9.8. Unit-Type House.

eastern room. Light from the rising winter solstice sun enters one of these ports and falls in the room's northwestern corner near a protruding wall. At summer solstice, another port allows a shaft of light to fall into the room's southwestern corner. This effect occurs nearly an hour after sunrise because a giant boulder prevents rays from the rising sun from reaching

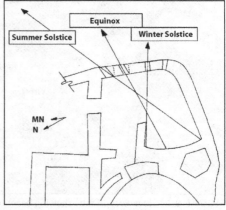

FIGURE 9.9. Unit-Type House showing sight-lines to the solstices and equinox.

the port. At the equinoxes, an area of the wall halfway between the low extending wall and the southwestern corner is illuminated at sunrise. Frank Hamilton Cushing reported that ordinary Zuni families kept tabs on the accuracy of the sun priest's predictions by means of portholes in the walls of their homes.[4] Perhaps the solar events commemorated at Unit-Type House were designed by such a family to establish their own personal communication with the sun and the powers it represented.

Cajon Group

The complex set of structures of Cajon Group lies about 6 miles southwest of Hovenweep Castle (Figures 9.10 and 9.11). The western wall of a tower room (A) is pierced by three portholes, and these may have func-

tioned in a manner similar to both Hovenweep Castle and Unit-Type House. Two of the portholes appear to have been associated with winter and summer solstice sunset, and the other is oriented to the equinox sunsets. As is the case with the other portholes, there are no special features or markings where the beams strike

FIGURE 9.10. Western building (B) in the Cajon Group.

FIGURE 9.11. Cutthroat Castle: Built for defense, overlooking a small reservoir.

the inside wall, and it is impossible to confirm whether these ports were intentionally devised as calendars. At winter solstice, rays from the setting sun probably would have been blocked by the nearby building (B). But the light's movement along the wall prior to and following the equinoxes would have allowed anticipation of the events.

The tower building may have an additional calendrical function in relation to building B to the west.[5] Shortly after fall equinox, the western building casts a shadow over part of the tower building, with the shadow moving northward until it covers only the western wall of the tower at winter solstice sunset. The shadow then reverses its path until the spring equinox, when both buildings are un-shadowed. Shortly afterward, the tower building begins to cast a shadow over its western counterpart at sunrise. The shadow effect appears to continue on the western building until some time near the summer solstice, but the ruined state of the tower building prevents determination of the exact date. Although we cannot be certain that the residents deliberately constructed the two buildings to serve as a giant seasonal sundial, they should have noticed the coincidence of the shadowing effects with the seasons and perhaps made use of it for honoring the sun and its cycles.

Holly House

On a rock panel about 300 feet to the south of Holly House, just below the rim of the canyon, residents of the area had pecked two spirals and a set of three concentric circles with a dot in the center. There also is a wavy line that may represent a plumed serpent and two connected circles, each with a dot in the center. About forty-five minutes after local sunrise, a horizontal spear of light enters the panel from the left and cuts across the left-hand spiral (Figure 9.13). Soon a second horizontal sliver of light enters from the right, passing below the center of the concentric circles, and shortly the two shafts of light join. They broaden and move downward across the petroglyph panel as the sun rises (Figure 9.14). The entire process takes only seven minutes, and, in its quiet certitude, it is spellbinding.

As was the case at Fajada Butte, the Ancestral Puebloans did not move the huge boulders into place to achieve the required play of light along the carvings. Rather, someone noticed the unique patterns formed by the rising sun at summer solstice and pecked the designs accordingly. The solar

FIGURE 9.12. Holly House.

FIGURE 9.13. Sun dagger near Holly House. As the sun rises, the single shaft of light coming from the left, near the spiral, is joined by a second shaft emerging from the right.

FIGURE 9.14. The merged and broadened shafts of sunlight.

FIGURE 9.15.
Holly Tower.

sites at Hovenweep, like those at Chaco Canyon, testify to the skill and ingenuity of the Ancestral Puebloans in utilizing natural phenomena at Holly House and in engineering novel effects of light and shadow in extraordinarily dangerous times.

Castle Rock, Sand Canyon, and Goodman Point Pueblos

These three Pueblo villages lie some fifteen miles to the east of Hovenweep. All had been the targets of attacks, and all were abandoned approximately between 1080 and 1085.[6] Castle Rock had been built soon after 1256, around a small butte. Its approximately fifteen households may have hoped that the butte would have provided protection. The village contained some thirty-seven rooms, sixteen kivas, and nine towers. The last tree ring date is 1274, and it was destroyed in an attack sometime between 1280 and 1285. The violence of the attack is extraordinary. At least forty-one of the inhabitants died in the attack. Human remains were found lying on floors, in collapsed roofs, and some on top of the butte. Some of the victims were subject to torture, decapitation, dismemberment, scalping, face removal, and cannibalism. Children and women were not spared. Among the dead were a forty-year-old woman and an eight-year-old child who had been killed by one or more blows to the back of the head, possibly while attempting to flee. The defensive nature of many of the buildings and towers of Hovenweep can be understood in the context of such a threat.

The villages of Sand Canyon and Goodman Point had a different defensive strategy from that at Hovenweep. Each was built around a spring and surrounded on three sides by a defensive wall. Sand Canyon contained ninety kivas and fourteen towers, and was occupied between 1250 and 1280 with a population of 225 to 500 people. Judging from healed bone fractures of some of its inhabitants, Sand Canyon had survived one or more attacks, and was probably abandoned close in time to the destruction of Castle Rock. Recent excavations at Goodman Point indicate that it, too, experienced attacks similar to those of Castle Rock and Sand Canyon.

What seems extraordinary about the Sand Canyon and Goodman Point communities is the attention they paid to the heavens during these stressful decades. Great kivas were constructed, aligned approximately to

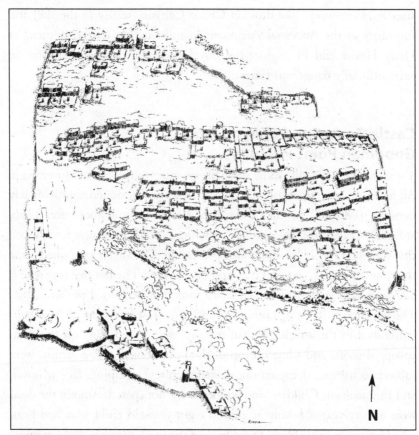

FIGURE 9. 16. Artist's reconstruction of Goodman Point. The towers in the center left have not been verified by excavation.

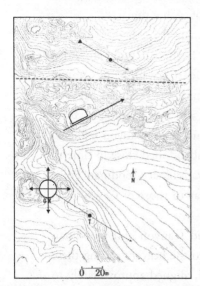

the cardinal directions, at Goodman Point and Sand Canyon. Both villages built D-shaped bi-wall structures with approximately the same orientation, such that their south walls were approximately oriented toward June solstice sunrise.

In 1988, Jean Kindig, an artist-archaeologist on our team, created a reconstruction of

FIGURE 9.17 Goodman Point. Note the cardinally oriented great kiva; a sight line to December solstice sunrise from the great kiva to a feature that is either a tower or a kiva; the D-shaped structure recently uncovered by Crow Canyon, which is aligned to June solstice sunrise; the 140-meter-long north wall containing some 30 kivas and 90 rooms; an alignment to December solstice parallel to the northeast portion of the site enclosing wall.

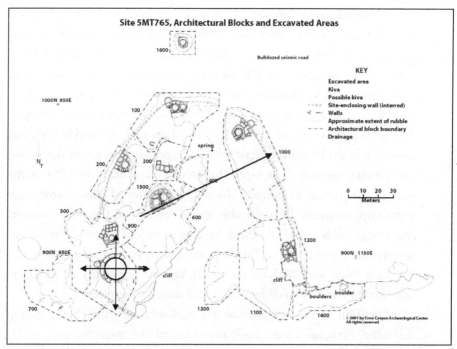

FIGURE 9.18. Sand Canyon Pueblo. Note the cardinally aligned great kiva and the June solstice alignment of the D-shaped bi-walled structure. Note the remarkable similarity to Figure 9.17.

the village, providing a valuable insight into its design: wrapped around the central spring, surrounded by defensive walls, with a great kiva in the south-west.[7] Her reconstruction was based upon a map made by Crow Canyon Archaeological Center, an aerial survey map made by Mesa Verde National Park, and our theodolite measurements (Figures 9.16 and 9.17). Recent excavations by the Crow Canyon Archaeological Center reveal that two of the towers near the spring were actually kivas, and also the presence of a D-shaped structure nearly identical in size and orientation to the one at Sand Canyon. The Crow Canyon archaeologists have identified 110 kivas. Goodman Point was occupied during the same period as Sand Canyon, between 1250 and 1280, and had a slightly larger population of 500 to 700 people.

Summary

In the latter half of the thirteenth century, life was getting very difficult throughout the Southwest. The climate was getting dryer and colder, and crops were failing. Famine and competition for scarce resources led to social instability and violence, which in some cases became excessive. The isolated towers of Hovenweep were one response to these threats. Another approach was followed by the villages of Sand Canyon and Goodman Point, which were built in defensible locations next to water sources and were surrounded by protective walls. Isolated homesteads in the neighborhood were abandoned as people aggregated into these more secure villages. The large community of Yellow Jacket had no such defensible wall, but towers built on its northeastern edge were probably to guard the spring and protect residents. The Great Tower was constructed in 1254 at the start of this turbulent period. The cliff enclosures of Mesa Verde were superb opportunities for defensible spaces. Some had their own springs; many were hidden, and all were defensible to various degrees. The tops of inaccessible buttes, such as Fajada Butte and other pinnacles sites in the San Juan Basin and the Kayenta region, became temporary refuges, perhaps for the most vulnerable members of the communities.

Attention to the heavens was a remarkably common theme during this time. Structures were built that marked the solstices, such as D-shaped bi-walls at Sand Canyon and Goodman Point and the building in the Cajon Group. Portholes in walls at several buildings in Hovenweep established the solstices and equinoxes. Light and shadows marked the solstices on petroglyphs at Fajada Butte and near Holly House. Major investments in labor were necessary to produce the great kivas and D-shaped buildings. While defense from their enemies would have been their major concern, people still turned their attention to the heavens.

CHAPTER 10

EPILOGUE

The chronological development of astronomy in the Four Corners area between 700 and 1300 CE is the story of the responses of a relatively isolated culture to a variety of needs and influences. By contrast, in northern Europe the fundamental sources of astronomy were not local. Astronomical information came from a vast variety of other sources— Italy, Greece, Mesopotamia, and even India. It was carried by writing, mathematics, and visiting scholars. The residents of Chaco Canyon had trade contacts covering a large area, but there is no persuasive evidence for the transfer of substantial astronomical information from outside the San Juan Basin. Puebloan astronomy was probably largely home-grown and is certainly a testament to the clear thinking and sharp eyes of ancient sky watchers.

The history of Chacoan astronomy appears to start north of the San Juan River in the 700s, where living quarters and ceremonial spaces were constructed approximately parallel to the north-south axis of the heavens. Perhaps this tradition reveals deep cultural roots reaching to nomadic ancestors who needed to establish north during their migrations. The interest in incorporating the north-south meridian in residential and ritual architecture was carried by migrants to Chaco Canyon sometime between 850 and 900 CE. It is possible that the appearance in 1006 of the most brilliant supernova ever seen by humans amplified the meaning and power of the north-south meridian that crossed Chaco Canyon. Sometime in the eleventh century, the technique of using shadow casting by a gnomon to determine true north was invented, and by 1085 it was employed to achieve north-south alignments in great houses and great kivas with remarkable precision. The concept of a horizon calendar must also have been invented early in that century, which enabled the leaders in the canyon to schedule system-wide periodic festivals. During the eleventh century, lunar standstills were discovered at Chimney Rock, becoming another element of sacred and cyclical time. A second wave of astronomical traditions began after the demise of the Chacoan system around 1125 CE. The new astronomy may have been more related to features of the sacred landscape and devotion to the sun and moon, such as what we find in Mesa Verde and Yellow Jacket. Finally in the latter half of the thirteenth century, when the climate deteriorated and conditions for dry-land farming became increasingly difficult, peace was shattered and astronomy must have taken on a different meaning than it had during Pax Chaco. In the towers of Hovenweep, the defensive villages of Sand Canyon and Goodman Point, and the inaccessible pinnacle of Fajada Butte, we find indications that people turned to sky gods for help and protection during those turbulent and dangerous times.

There are still great puzzles and mysteries remaining in our exploration of the prehistoric astronomy in the canyons and mesas of the Southwest.

1. First, how did these prehistoric inhabitants of the Southwest acquire this astronomical knowledge? Were there just a few "astronomer-priests" who knew about things like solstices and lunar standstills, or were they common knowledge for people who lived outdoors much more than we do today?

2. Why did they put so much effort into astronomy? Monumental ceremonial structures like the Great Kiva of Casa Rinconada in Chaco, the Great House of Pueblo Bonito, and the Sun Temple of Mesa Verde were designed around astronomy.

3. Finally, what astronomy have we missed? What is still out there? What light and shadow effects on ruined buildings and petroglyphs on canyon walls have we not yet seen or comprehended? This is the great adventure that awaits the curious visitor. Once you know what to look for, where the sun rises and sets, you can participate in the process of discovery.

When visiting these ruins, please treat them with respect, for modern Puebloan people consider that they have never been abandoned.

NOTES

CHAPTER 1: The Ancestral Puebloan Astronomer

1. Frank Waters, *The Book of the Hopi* (New York: Viking Press, 1963), pp. 8–9.

2. John Hemming, *The Conquest of the Inca* (London: Harcourt, 1970), pp. 172–173.

3. J. McKim Malville, Romauld Schild, Fred Wendorf, and Robert Brenmer. "Astronomy of Nabta Playa," in *African Cultural Astronomy,* eds.. Jarita Holbrook, Rodney Medupe, and Johnson Urama (Springer, 2008), pp. 131–143.

4. George Michell, *The Hindu Temple* (Chicago: The University of Chicago Press, 1988); J. McKim Malville and R. N. Swaminathan, "People, Planets, and the Sun: Surya Puja in Tamil Nadu, South India, *Culture and Cosmos* Vol. 2, no 1 (1998): 3–15.

5. Johanna Broda, David Carrasco, and Eduardo Matos Moctezuma, *The Great Temple of Tenochtitlán: Center and Periphery in the Aztec World* (Berkeley: University of California Press, 1987).

6. William Lipe, "Notes from the North," in *The Archaeology of Chaco Canyon,* ed. Stephen Lekson (Santa Fe: School of American Research, 2006), p. 268.

7. J. McKim Malville, "Astronomy and Social Integration Among the Ancestral Pueblo," *Proceedings of the Ancestral Pueblo Symposium,* 1991, ed. Art Hutchinson and Jack E. Smith (Mesa Verde Museum Association, 1993); W. James Judge, "Chaco: Current Views of Prehistory and the Regional System," in *Chaco and Hohokam,* ed. Patricia L. Crown and W. James Judge (Santa Fe: School of American Research Press 1991), pp. 11–30.

8. J. McKim Malville, Frank Eddy, and Carol Ambruster, "Lunar Standstills at Chimney Rock," *Archaeoastronomy,* Supplement to the *Journal for the History of Science* 16 (1991): 543–550; J. McKim Malville, "Ceremony and Astronomy at Chimney Rock," in *Chimney Rock, The Ultimate Outlier,* ed. J. McKim Malville (Lanham: Lexington Books, 2004).

CHAPTER 2: The Roots of Astronomy

1. William Lipe, "Notes from the North," in *The Archaeology of Chaco Canyon,* ed. Stephen Lekson (Santa Fe: School of American Research, 2006).

2. E. James Dixon, *Bones, Boats, and Bison* (Albuquerque: University of New Mexico Press, 1999).

3. Richard H. Wilshusen and Ruth Van Dyke, "Chaco's Beginnings," in *The Archaeology of Chaco Canyon,* ed. Stephen Lekson (Santa Fe: School of American Research, 2006).

4. W. James Judge, "Chaco Canyon-San Juan Basin," in *Dynamics of Southwest Prehistory,* ed. Linda Cordell and George Gummerman (Washington, D.C.: Smithsonian Institution Press, 1989): 241; J. McKim Malville and Nancy J. Malville, "Pilgrimage and Periodic Festivals as Processes of Social Integration in Chaco Canyon," *Kiva*: 66: 327–344 (2001).

5. Kristin Kucklman, Ricky Lightfoot, and Debra Martin, "The Bioarchaeology and Taphonomy of Violence at Castle Rock and Sand Canyon Pueblos, Southwestern Colorado," *American Antiquity* 67 (2002): 486–515.

6. Timothy A. Kohler and Kathryn Kramer, "Raiding for Women in the Pre-Hispanic Northern Pueblo Southwest?" *Current Anthropology* 47 (2006): 1035–1045.

7. Dabney Ford, "Architecture on Fajada Butte," in *The Spadefoot Toad Site,* ed. Thomas C. Windes (Santa Fe: Branch of Cultural Research, Division of Anthropology, National Park Service, 1993).

8. Jonathan Haas and Winifred Creamer, "The Role of Warfare in the Pueblo III World," in *The Prehistoric Pueblo World, A.D. 1150–1350,* ed. Michael Adler (Tucson: University of Arizona Press, 2000): 205–213.

9. John Stein and Andrew Fowler, "Looking Beyond Chaco in the San Juan Basin and Its Peripheries," ed. Michael A. Adler (Tucson: The University of Arizona Press, 1966): 120.

10. Jonathan Hass and Winifred Creamer, "The Role of Warfare in the Pueblo III World," op. cit. p. 208.

CHAPTER 3: The Dome of the Sky

1. Linda S. Cordell, *Prehistory of the Southwest* (New York: Academic Press, 1984).

2. Matthew W. Stirling, "Origin Myth of Acoma," *Bureau of American Ethnology Bulletin* 135: 1–123.

3. Ray A. Williamson, "Light and Shadow, Ritual, and Astronomy in Ancestral Pueblo Structures," in *Astronomy and Ceremony in the Prehistoric Southwest,* ed. John B. Carlson and W. James Judge. Papers of the Maxwell Museum of Anthropology, Number 2, Albuquerque, 1983, pp. 99–119.

4. Linda S. Cordell, *Prehistory of the Southwest,* p. 254.

5. Stephen H. Lekson, *Great Pueblo Architecture of Chaco Canyon* (Albuquerque: University of New Mexico Press, 1986).

6. Ray A. Williamson, "Casa Rinconada: A Twelfth-Century Ancestral Pueblo Kiva," in *Archaeoastronomy in the New World,* ed. Anthony F. Aveni (Cambridge: Cambridge University Press, 1982), pp. 205–219.

7. Michael Zeilik, "The Ethnoastronomy of the Historic Pueblos: Calendrical Sun Watching," *Archaeoastronomy,* Supplement to the *Journal for the History of Astronomy* 8 (1985): 1–24.

8. Florence H. Ellis, "A Thousand Years of the Pueblo Sun-Moon-Star Calendar," in *Archaeoastronomy in Precolumbian America,* ed. Anthony F. Aveni (Austin: University of Texas, 1975), pp. 59–87.

9. Ray Norris, "Megalithic Observatories in Britain: Real or Imagined?" in *Records in Stone, Papers in Memory of Alexander Thom,* ed. C. L. N. Ruggles (Cambridge: Cambridge University Press, 1988), pp. 262–276.

10. A. Sofaer, A. V. Zinser, and R. M. Sinclair, "A Unique Solar Marking Construct," *Science* 206 (1979): 283–291; A. Sofaer, R. M. Sinclair, and L. E. Doggett, "Lunar Markings on Fajada Butte, Chaco Canyon, New Mexico," in *Archaeoastronomy in the New World,* ed. Anthony F. Aveni (Cambridge: Cambridge University Press, 1982), pp. 169–181.

11. Anthony F. Aveni, "Archaeoastronomy in the Southwestern United States: A Neighbor's Eye View," in *Astronomy and Ceremony in the Prehistoric Southwest,* ed. J. B. Carlson and W. James Judge (Albuquerque: Maxwell Museum, 1987), pp. 9–23; John B. Carlson, "Romancing the Stone, or Moonshine on the Sun Dagger," in *Astronomy and Ceremony in the Prehistoric Southwest,* ed. J. B. Carlson and W. James Judge (Albuquerque: Maxwell Museum, 1987), pp. 71–88; Michael Zeilik, "A Reassessment of the Fajada Butte Solar Marker," *Archaeoastronomy,* Supplement to the *Journal for the History of Astronomy* 16 (1985): S69–S85.

CHAPTER 4: Sky Watchers

1. Michael Zeilik, "Sun Shrines and Sun Symbols in the U.S. Southwest," *Archaeoastronomy,* Supplement to the *Journal for the History of Astronomy* 16 (1985): S86–S96; Michael Zeilik, "The Ethnoastronomy of the Historic Pueblos, I: Calendrical Sun Watching," *Archaeoastronomy,* Supplement to the *Journal for the History of Astronomy* 16 (1985): S1–S24.

2. Frank Hamilton Cushing, "My Adventures in Zuni," in *Zuni: The Selected Writings of Frank Hamilton Cushing,* ed. J. Green (Lincoln: University of Nebraska Press, 1979), pp. 116–117.

3. Michael Zeilik, "The Ethnoastronomy of the Historic Pueblos, I: Calendrical Sun Watching."

4. Stephen C. McCluskey, "Historical Archaeoastronomy: The Hopi Example," in *Archaeoastronomy in the New World,* ed. Anthony F. Aveni (Cambridge: Cambridge University Press, 1982), pp. 31–57.

5. Michael Zeilik, "The Ethnoastronomy of the Historical Pueblos, I: Calendrical Sun Watching."

6. Alfonso Ortiz, *The Tewa World: Space Time, Being, and Becoming in a Pueblo Society* (Chicago: University of Chicago Press, 1969).

7. Michael Zeilik, "The Ethnoastronomy of the Historic Pueblos, II: Moon Watching," *Archaeoastronomy,* Supplement to the *Journal for the History of Astronomy* 17 (1986): S1–S22.

8. Florence H. Ellis, "A Thousand Years of the Pueblo Sun-Moon-Star Calendar," in *Archaeoastronomy in Precolumbian America,* ed. Anthony F. Aveni (Austin: University of Texas, 1975), pp. 59–87.

9. Michael Zeilik, "The Ethnoastronomy of the Historic Pueblos, II: Moon Watching."

10. Barbara Tedlock, "Zuni Sacred Theater," *American Indian Quarterly* 7: 93–109.

11. Michael Zeilik, "The Ethnoastronomy of the Historic Pueblos, II: Moon Watching."

12. Ibid.

13. Barbara Tedlock, "Zuni Sacred Theater"; Michael Zeilik, "The Ethnoastronomy of the Historic Pueblos, II: Moon Watching."

14. Florence Hawley Ellis, "A Thousand Years of the Pueblo Sun-Moon-Star Calendar."

15. Michael Zeilik, "The Ethnoastronomy of the Historic Pueblos, II: Moon Watching."

CHAPTER 5: Chaco Canyon

1. David G. Noble, ed., *New Light on Chaco Canyon* (Santa Fe: School of American Research, 1984); David G. Nobel, ed *The Enigma of Chaco* (Santa Fe: School of American Research, 2004).

2. Chris Kincaid, ed., *Chaco Roads Project,* Phase I (Santa Fe: Bureau of Land Management, 1983); Stephen Lekson, *Great Pueblo Architecture of Chaco Canyon* (Albuquerque: University of New Mexico Press, 1986).

3. David Clark and Richard Stephenson. *The Historical Supernovae* (Oxford: Pergamon Press, 1977).

4. Stephen Lekson. *The Chacoan Meridian: Centers of Political Power in the Ancient Southwest.* (Walnut Creek: AltaMira Press, 1999).

5. Stephen Lekson, Thomas Windes, and Peter McKenna, "Architecture," in *The Archaeology of Chaco Canyon: An Eleventh-Century Pueblo Regional Center* (Santa Fe: School of American Research Press, 2006); W. H. Wills and Thomas Windes, "Evidence for Population Aggregation and dispersal during the Basketmaker III Period in Chaco Canyon, New Mexico." *American Antiquity* 54 (1989): 347–369.

6. Ibid.

7. Anna Sofaer. *Chaco Astronomy* (Santa Fe: Ocean Tree Books, 2008), pp. 81–113.

8. John C. Brandt, S. P. Maran, R. A. Williamson, R. S. Harrington, C. Cochran, M. Kennedy, W. J. Kennedy, and V. D. Chamberlain, "Possible Rock Art Records of the Crab Nebula Supernova in the Western United States," *Archaeoastronomy in Pre-Columbian America,* ed. Anthony F. Aveni (Austin: University of Texas Press, 1975), pp. 45–57.

9. Frank Hamilton Cushing, "My Adventures in Zuni," in *Zuni: Selected Writings of Frank Hamilton Cushing,* ed. J. Green (Lincoln: University of Nebraska Press, 1979), pp. 116–117.

10. Florence H. Ellis, "A Thousand Years of the Pueblo Sun-Moon-Star Calendar," in *Archaeoastronomy in Precolumbian America,* ed. Anthony F. Aveni (Austin: University of Texas, 1975), pp. 59–87.

11. Ray A. Williamson, "Casa Rinconada: A Twelfth-Century Ancestral Pueblo Kiva," in *Archaeoastronomy in the New World,* ed. Anthony F. Aveni (Cambridge: Cambridge University Press, 1982), pp. 205–219.

12. Ray A. Williamson, *Living the Sky* (Boston: Houghton Mifflin, 1984).

13. Thomas C. Windes, "Stone Circles of Chaco Canyon," Reports of the Chaco Center No. 5 (Albuquerque: Division of Chaco Research, National Park Service, 1978).

14. Anna Sofaer, *Chaco Astronomy,* op. cit. pp. 23–27.

15. Dabney Ford, "Architecture on Fajada Butte," in *The Spadefoot Toad Site: Investigations at 29SJ 629 Chaco Canyon, New Mexico*, Vol. 1, ed. Thomas Windes (Santa Fe: Branch of Cultural Research, Division of Anthropology, National Park Service, 1993), pp. 471–482.

16. Ibid.

17. Anna Sofaer, *Chaco Astronomy,* op. cit. pp. 49–72.

18. Anna Sofaer, *Chaco Astronomy,* op. cit. pp. 39–48.

19. Ibid.

20. Michael Zeilik, "The Fajada Butte Solar Marker: A Reevaluation," *Science* 228 (1985): 1311–1313.

CHAPTER 6: Chimney Rock

1. R. P. Powers, W. B. Gillespie, and S. H. Lekson, *The Outlier Survey: A Regional View of Settlement in the San Juan Basin* (Albuquerque: Division of Cultural Research, National Park Service, 1983).

2. F. W. Eddy, *Archaeological Investigations at Chimney Rock Mesa: 1970–1972* (Boulder: Colorado Archaeological Society, 1977); I. D. Webster, *An Archaeological Survey of the West Rim of the Piedra River* (Durango, CO: San Juan National Forest, 1983); F. W. Eddy, "Past and Present Research at Chimney Rock" in *Chimney Rock: The Ultimate Outlier*, ed. J. McKim Malville (Lanham: Lexington Books 2004), pp.23–50.

3. F. H. Ellis, "Differential Pueblo Specialization in Fetishes and Shrines," *Anales* 1967–1968, Sobretiro, Septima epoca, Torno I, Mexico; F. H. Ellis and J. J. Brody, "Ceramic Stratigraphy and Tribal History at Taos Pueblo," *American Antiquity* 29 (1964): 3.

4. S. W. Lekson, *Great Pueblo Architecture of Chaco Canyon* (Albuquerque: University of New Mexico Press, 1984).

5. R. G. Vivian, "An Inquiry into Prehistoric Social Organization in Chaco Canyon, New Mexico," in *Reconstructing Prehistoric Pueblo Societies,* ed. W. A. Longacre (Albuquerque: University of New Mexico Press, 1970), pp. 59–83.

6. J. McKim Malville, "Ceremony and Astronomy at Chimney Rock," op. cit.

7. Gary Fairchild, J. McKim Malville, and Nancy Malville, "Chimney Rock as a Ceremonial Center and Port-of-Trade within the Chaco System," *Viewing the Sky Through Past and Present Cultures: Selected Papers from the Oxford VII Conference on Archaeoastronomy,* ed. Todd Bostwick and Bryan Bates (Phoenix: Pueblo Grande Museum Anthropological Papers no. 15, 2006) pp. 259–274.

8. Michael Zeilik, "The Ethnoastronomy of the Historic Pueblos, II: Moon Watching," *Archaeoastronomy,* Supplement to the *Journal for the History of Astronomy* 10 (1986): S1–S22.

9. F. H. Ellis and L. Hammack, "The Inner Sanctum of Feather Cave, a Mogollon Sun and Earth Shrine Linking Mexico and the Southwest," *American Antiquity* 30 (1968): 25.

10. L. D. Webster, "An Archaeological Survey of the West Rim of the Piedra River"; J. McKim Malville, "Ceremony and Astronomy at Chimney Rock," *Chimney Rock: The Ultimate Outlier*, ed. J. McKim Malville (Lanham: Lexington Book 2004), pp.131–150.

CHAPTER 7: Yellow Jacket

1. Kristin Kuckelman, *The Archaeology of Yellow Jacket Pueblo*, http://www.crow-canyon.org/YellowJacket.htm

2. Paul Wheatley, *The Pivot of the Four Quarters* (Chicago: Aldine Publishing, 1971).

3. Frederick Lange, Nancy Mahaney, Joe Ben Wheat, and Mark L. Chenault, *Yellow Jacket: A Four Corners Ancestral Pueblo Ceremonial Center* (Boulder: Johnson Books, 1986).

4. M. Zeilik, "Sun Shrines and Sun Symbols in the U.S. Southwest," *Archaeoastronomy*, Supplement to the *Journal for the History of Astronomy* 15 (1985): 86–96.

5. Edgar Hewett, *Pajarito Plateau and Its Ancestral People* (Albuquerque: University of New Mexico Press, 1938).

6. F. H. Ellis and L. Hammack, "The Inner Sanctum of Feather Cave, a Mogollon Sun and Earth Shrine Linking Mexico and the Southwest," *American Antiquity* 33 (1968): 25–44.

7. Doris Heyden, "Caves, Gods, and Myths: World-View and Planning in Teotihuacán," in *Mesoamerican Sites and World Views,* ed. Elizabeth Benson (Washington, DC: Dumbarton Oaks, 1981).

8. Broda, Johanna, "Templo Mayor as Ritual Space," in *The Great Temple of Tenochtitlán* (Berkeley: University of California, 1987), pp. 61–123.

9. Florence H. Ellis, "A Thousand Years of the Pueblo Sun-Moon-Star Calendar," in *Archaeoastronomy in Precolumbian America,* ed. Anthony F. Aveni (Austin: University of Texas, 1975), pp. 59–87.

10. J. McKim Malville, "The Cosmic and the Sacred at Mesa Verde and Yellow Jacket," in *The Mesa Verde World: Explorations in Ancestral Pueblo Archaeology,* ed. David Noble (Santa Fe: School of American Research, 2006). pp. 84–91.

11. Ibid.

CHAPTER 8: Mesa Verde

1. J. W. Fewkes, "A Sun Temple in the Mesa Verde National Park," *Art and Archaeology* 3: 341–346 (1916); J. W.Fewkes, *Excavation and Repair of Sun Temple, Mesa Verde National Park* (Washington, DC: Department of the Interior, 1916); *Explorations and Field-Work of the Smithsonian Institution in 1920.* Miscellaneous Collections, vol. 72, no. 6. (Washington, D.C.: Smithsonian Institution, 1921).

2. A. H. Rohn, *Cultural Change and Continuity on Chapin Mesa* (Lawrence, KS: Regents Press, 1977).

3. J. McKim Malville, "Astronomy and Social Integration among the Ancestral Pueblo," *Proceedings of the Anasazi Symposium,* 1991, ed. A. Hutchinson and J. E. Smith (Mesa Verde: Mesa Verde Museum Association, 1993).

4. Michael Zeilik, "Keeping the sacred and planting calendar: archaeoastronomy in the Pueblo Southwest," *World Archaeoastronomy,* ed. Anthony F. Aveni (Cambridge: Cambridge University Press, 1989) pp. 143–166.

5. J. McKim Malville and Gregory Munson, "Pecked Basins of the Mesa Verde, *Southwestern Lore* 64:1–35, 1998.

6. Ibid.

7. Ibid.

CHAPTER 9: The Towers and Castles of Hovenweep

1. Kristin A. Kuckelman, "Ancient Violence in the Mesa Verde Region," and Mark D. Varian, "Turbulent Times in the Mesa Verde World," in *The Mesa Verde World,* ed. David G. Noble (Santa Fe: School of American Research, 2006), pp. 127–135 and pp. 39–47 respectively.

2. Joseph C. Winter, "Hovenweep through Time," in *Exploration,* ed. David G. Noble (Santa Fe: School of American Research, 1985), pp. 22–28.

3. Ray A. Williamson, Howard J. Fisher, and Donnel O'Flynn, "Ancestral Pueblo Solar Observatories," in *Native American Astronomy,* ed. Anthony F. Aveni (Austin: University of Texas Press, 1977), pp. 203–217; Michael Zeilik, "Anticipation in Ceremony: The Readiness Is All," in *Astronomy and Ceremony in the Prehistoric Southwest,* Papers of the Maxwell Museum of Anthropology, Number 2, Albuquerque 1987; Ray A. Williamson, "Light and Shadow, Ritual and Astronomy in Ancestral Pueblo Structures," in *Astronomy and Ceremony in the Prehistoric Southwest,* Papers of the Maxwell Museum of Anthropology, Number 2, Albuquerque, 1986, pp. 99–119.

4. Frank Hamilton Cushing, "My Adventures in Zuni," in *Zuni: The Selected Writings of Frank Hamilton Cushing,* ed. J. Green (Lincoln: University of Nebraska Press, 1979), pp. 116–117.

5. Ray A. Williamson, "Light and Shadow, Ritual and Astronomy in Ancestral Pueblo Structures."

6. Kristin Kuckelman, Ricky Lightfoot and Debora Martin, "The Bioarchaeology and Taphonomy of Violence at Castle Rock and Sand Canyon Pueblos, Southwestern Colorado." *American Antiquity* 67: 486–513, 2002; Kristin Kuckelman and Grant Coffey, 2007 Report of 2006 Research at Goodman Point Pueblo (5MT604) Montezuma County, Colorado. www.crocanyon.org/goodmanpoint2006.

7. John McKim Malville and James Walton, "Organization of large settlements of the northern Anasazi," in *Archaeoastronomy in the 1990s,* ed. Clive Ruggles (Loughborough, UK: Group D Publications 1993) pp. 242–250.

CREDITS

CHAPTER 1
1.1. Malville
1.2. Malville
1.3. Malville
1.4. Malville
1.5. Malville
1.6. Malville
1.7. Codex Ixtlilixochtl, Bibliothéque National, Paris
1.8. George DeLange
1.9. Helen Richardson
1.10. NASA
1.11. Australian National Observatory

CHAPTER 2
2.1. After Figure 7.1, Richard H. Wilshusen and Ruth Van Dyke, 2006, "Chaco's Beginnings," in ed. Stephen H. Lekson, *The Archaeology of Chaco Canyon: An Eleventh-Century Pueblo Regional Center* (Santa Fe: School of American Research)
2.2. Anasazi Heritage Center, Bureau of Land Management.
2.3. After Prudden 1903: 235
2.4. After Figure 1, Richard H. Wilshusen, 1989, "Unstuffing the Estufa: Ritual Floor Features in Anasazi Pit Structures and Pueblo Kivas," in eds. W. D. Lipe and Michelle Hegmon, *The Architecture of Social Integration in Prehistoric Pueblos* (Cortez: Crow Canyon Archaeological Center).
2.5. Adapted from Joel M. Brisbane, Allen E. Kane, and James N. Norris, 1988, "Excavations at McPhee Pueblo (5MT4475), A Pueblo I and Early Pueblo II Multicomponent Village, in eds. A. E. Kane and C. K. Robinson, *Dolores Archaeological Program: Anasazi Communities at Dolores: McPhee Village,* pp. 63–401. (Denver: U.S. Bureau of Reclamation).

2.6. After Figure 4.1, Lekson 1999.
2.7. Robert Beehler
2.8. J. Q. Jacobs

CHAPTER 3
3.1. a. James Walton; b. Malville
3.4. Snowdon Hodges
3.7. William K. Hartman, Astronomy: The Cosmic Journey, Wadsworth Publishing Co., Belmont, California, 1978.
3.8. John Lubs, Griffith Observatory
3.13. After C. Daryll Forde, "Hopi Agriculture and Land Ownership," *Journal of the Royal Anthropological Institute of Great Britain and Ireland* 61, pp. 357–405, 1931.
3.14. After E. C. Krupp, *In Search of Ancient Astronomies,* ed. E. C. Krupp, McGraw-Hill, New York, 1978.
3.15. Malville
3.16. Frank Eddy

CHAPTER 5
5.1. National Park Service
5.2. David Wilcox
5.3. After Fig. 4.13: John R. Stein, Dabney Ford, and Richard Friedman, 2003, "Reconstructing Pueblo Bonito," in ed. Jill E. Neitzel, *Pueblo Bonito: Center of the Chacoan World* (Washington: Smithsonian Books).
5.4. Stephen Lekson
5.5. Tyler Nordgren
5.6. Tyler Nordgren
5.7. Richard Keen
5.8. Malville
5.9. Reading Museum
5.10. Malville
5.11. Malville

5.12. After Ray W. Williamson, "Casa Rinconada, A Twelfth Century Ancient Pueblo Kiva," in *Archaeoastronomy in the New World,* ed. A.F. Aveni, Cambridge University Press, Cambridge, 1982, pp. 205–219.

5.13. Malville

5.14. Malville

5.15. Malville

5.16. Malville

5.17. GB Cornucopia

5.18. Malville

5.19. a. Malville; b. Gugliemo Temple; c. NASA

5.20. Malville

5.21. GB Cornucopia

5.22. GB Cornucopia

5.23. Mike King

5.24. Malville

5.25. Malville

5.26. Adapted from National Park Service

5.27. GB Cornucopia

5.28. GB Cornucopia

5.29. a. National Park Service; b. High Altitude Observatory

CHAPTER 6

6.1. U.S. Forest Service

6.2. Robert Powell

6.3. Malville

6.4. Robert Powell

6.5. Chimney Rock Interpretative Association

6.8. Frank Eddy

6.9. Malville

6.10. Helen Richardson

6.11. Frank Eddy

6.12. Dale Lightfoot

6.13. Jean Jeançon

6.14. James Walton

6.15. Malville

6.16. Malville

6.18. Malville

6.19. Malville

6.20. Malville

6.21. Jean Kindig

6.23. Malville

CHAPTER 7

7.1. Jean Kindig

7.3. Malville

7.4. Malville

7.5. Malville

7.7. Malville

7.8. Rudy Poglitsch

7.9. Rudy Poglitsch

7.10. Jean Kindig

CHAPTER 8

8.1. Malville

8.2. Malville

8.4. Preston Fisher

8.5. Malville

8.7. Malville

8.12. Malville

CHAPTER 9

9.1. Malville

9.2. National Park Service

9.3. Malville

9.4. Malville

9.5. R. Williamson

9.6. R. Williamson

9.7. Malville

9.8. J. Q. Jacobs

9.9. R. Williamson

9.10. J. Q. Jacobs

9.11. J. Q. Jacobs

9.12. Malville

9.13. Malville

9.14. Malville

9.15. J. Q. Jacobs

9.16. Jean Kindig

9.17. Topographic map courtesy of Jack Smith, Mesa Verde National Park.

9.18. Crow Canyon Archaeological Center, 2004 The Sand Canyon Pueblo Database: http://www.crow-canyon.org/sandcanyondatabase. Date of use: 18 May 2008.

Index

A

Acoma people, 113
Altitude (celestial sphere),
    30-31, *30*, *31*
Ancestral Puebloan people, 43-48
    about, 130
    astronomy of, 3, 7, 11
    different from modern Puebloans, 43
    gnomons, 35
    kiva as astronomical symbol,
        26-29, *27*, *28*, *29*
    lunar standstills, 38-39, *40*, 41, *41*
    moon watching by, 46-48
    north-south orientation of
        buildings, 15, 33, 34, 35
    stargazing, 48
    sun watching by, 44-46
    sundials, 42
    winter solstice, 17, 21
Archaeoastronomy, 3
Astronomers, 3
Astronomy, dome of the sky, 25
Azimuth, 30, *30*
Aztec (Chaco Canyon), 53
Aztec people, Templo Mayor, 6, *6*, 113

B

Badger House, 133
Bandelier National Monument, 112
Basketmaker III site, 54
Bayeux tapestry, Halley's Comet in,
    60, 61
Big Dipper, 30
Birth legend of, 1
Bonfires, 51
Bonito Phase (Chaco Canyon),
    8, 10, 55, 56, 69, 76

C

Cajon Group, 136, 141-142, *141*, 148

Calendar
    traders and, 17, 18
    travelers and, 20
Calendar circle, in Sahara Desert,
    3-4, *3*, *4*, 15
Calendrical stations
    at Chaco Canyon, 21, 70, 79
    at Chimney Rock, 96
Casa Chiquito, 55
Casa Rinconada
    construction of, 55
    Great Kiva, *27*, 29, *29*, 60-64,
        60-63
    niches, 62-63
    north-south orientation of,
        8, 34, 56, 74, 79
    El Castellejo, 23
Castle Rock massacre, 12, 23, 145
Castle Rock Pueblo, 21, 145
Caves, as shrines, 112-113
Cedar Tree Tower,
    101, 123, *132*, 133-134
Celestial equator, 35-36
Celestial sphere, 25-26
    azimuth and altitude, 30, *30*, *31*, 32
    celestial equator, 35-36
    cyclic motion of the moon,
        38-39, *39*, *40*, 41
    cyclic motion of the sun, 36-38
    Greeks on, 25, 26-27
    kiva as astronomical symbol, 26-29
    path of the sun, 36
    Pole Star, 32-35
Ceremonial centers, at Yellow Jacket,
    102-103
Chac (deity), 113
Chaco Canyon, 49-79
    about, 49-50, 149-150
    astronomy of, 7-10, 50, 78-79,
        80, 149
    Bonito Phase, 8, 10, 55, 56, 69, 76
    buildings in, 50
    calendrical stations, 21, 70, 79